Optical Materials

Microstructuring Surfaces with Off-Electrode Plasma

Optical Materials
Microstructuring Surfaces with Off-Electrode Plasma

Nikolay L. Kazanskiy
Vsevolod A. Kolpakov

CRC Press
Taylor & Francis Group
Boca Raton London New York

CRC Press is an Imprint of the
Taylor & Francis Group, an **informa** business

CRC Press
Taylor & Francis Group
6000 Broken Sound Parkway NW, Suite 300
Boca Raton, FL 33487-2742

First issued in paperback 2019

© 2017 by Taylor & Francis Group, LLC
CRC Press is an imprint of Taylor & Francis Group, an Informa business

No claim to original U.S. Government works

ISBN-13: 978-1-138-19728-2 (hbk)
ISBN-13: 978-0-367-88626-4 (pbk)

**Visit the Taylor & Francis Web site at
http://www.taylorandfrancis.com**

**and the CRC Press Web site at
http://www.crcpress.com**

Contents

Foreword

Modern setups for vacuum plasma treatment of materials use reactors in which a gas discharge generates plasma in an electrode gap. The electrodes can be either inside or outside the vacuum chamber housing the reactor. This approach results in a more complex design and higher energy consumption by plasma sources, thus leading to higher costs. This can be explained by the following detrimental factors inherent in the existing methods of generating low-temperature plasma:

- An increase in the relative surface area being treated leads to a decrease in the etch rate, including through ion shadowing, neutral shadowing, and the microloading effect.
- There is a need to optimize system parameters.
- Plasma collides with reactor walls.
- The material, shape, and properties of the substrate surface affect the parameters of gas discharge.

These factors increase the complexity involved in the fabrication of micro- and nanostructures, increase costs, and lead to an unreasonable increase in the number of processing units, reactors, and process gases. The problem becomes even more acute for diffractive optics in terms of generating highly uniform microsized diffraction microreliefs on large-format wafers of various optical materials.

To solve this problem, this monograph proposes the use of high-voltage gas-discharge off-electrode plasma, a type of plasma that was first studied and tested at the Korolev State Aerospace University, Samara, Russia. The term *off-electrode plasma* is new, but the authors knowledgeably discuss it in detail, adequately describe its history, and support their discussion with references to authoritative literature.

I think that the topic raised in this monograph and the results it describes will be useful in various fields of science such as gas discharge physics, plasma physics, solid-state physics, and computer diffractive optics.

A. S. Bugayev
Academician
Professor
Head of the Department of Vacuum Electronics
Moscow Institute of Physics and Technology
(State University), Moscow, Russia

Preface

Surface microstructuring of optical materials is widely used in the manufacture of diffractive optical elements (DOEs) for various industries such as optical equipment (digital cameras, mobile phones, CD players, video projectors, etc.), medicine (laser systems, contact lenses, eye lenses), manufacturing (laser systems), communications (fiber optics), automotive (headlights with set beam patterns), and science (microinterferometers, scientific devices).

DOEs are transmissive or reflective thin-phase elements that operate on the principle of diffraction [1,2]. Most DOEs come as wafers transparent in the visible and infrared ranges and are made of materials that have a diffraction microrelief called *optical microrelief.* These materials include silicon dioxide, diamond-like films, polymers, and zinc selenide. Stepped (binary or multilevel) and continuous microreliefs have found the widest application. In this monograph, we propose and discuss methods for microstructuring the surfaces of optical materials to obtain binary microreliefs. The methods also provide the basis for creating multilevel microreliefs. Depending on the operating wavelength and refractive index of a given material, the profile height may vary from 0.5 to 7.5 µm and the zone width from 1 to 200 µm.

Surface microstructuring of optical materials is accomplished through (photo) lithography [3,4], use of bichromated gelatin [5] and liquid photopolymers [6], layered photoresist growth [7], direct laser ablation [8,9], direct laser writing [10], and vacuum plasma etching in high-frequency (HF) and superhigh-frequency (SHF) plasma [11,12]. Our analysis of studies related to this topic identified a need for drastically improving the microrelief quality of DOEs and the precision of microrelief parameters [9,13]; expanding the range of DOEs, including by fabricating highly uniform, variously shaped precision microreliefs on large-aperture wafers of various optical materials [1,2]; and reducing the inclination of microrelief-profile walls from the vertical [1,2]. The aspect ratio of trenches in diffraction microreliefs needs to be increased [14,15].

To prepare the templates and microreliefs discussed in this monograph, a broad range of specialized equipment (listed in References 1 and 2) is used. For precision recording of pattern layouts with diameters of up to 200 mm, we used the CLWS-200 laser-writing system and other equipment at Nanophotonics and Diffractive Optics, a shared center set up by the Korolev State Aerospace University and the Image Processing Systems Institute at the Russian Academy of Sciences.

The desired optical-microstructure quality is achieved through the use of plasma techniques whose key operations are surface cleaning, cleanliness measurement, application of a mask durably resistant to low-temperature plasma, substrate etching, and complete mask removal after etching.

Precision surface cleaning, high adhesion between the mask and the substrate surface, use of directed low-temperature plasma fluxes for uniform, anisotropic (directed) etching of the surface, and complete removal of the mask after etching make it possible to satisfy the quality requirements.

Precision cleaning is achieved through final cleaning of substrate surfaces and measuring surface cleanliness. There are multiple methods and tools for measuring surface cleanliness [16–23]. But because of their high cost and low productivity, many existing methods are unsuitable for measuring the surface cleanliness of large-format substrates when you need to quickly and accurately check whether the surface meets the process requirements. In addition, special operations—cleaning the indenter probe's surface and using substrate surfaces with reference contamination for calibration—are required for measurement. The existing methods have some other shortcomings as well: only a specific type of contamination at the surface under study can be measured, the surface gets soiled in the process, and the instrument readings are unstable.

The methods of immersion, condensation, isotopes, secondary ion mass spectrometry, Auger electron spectroscopy, neutron activation analysis, Rutherford backscattering spectrometry, scanning electron microscopy, electron spectroscopy for chemical analysis, and the like offer measurement accuracies of 10^{-6}–10^{-10} g/cm^2 [18,21,22]. But these methods are analytical, and because they are time-consuming and expensive, they are unsuitable for quick measurements in developing and using new methods for final cleaning of substrate surfaces in low-temperature plasma. None of the methods allows surface cleanliness to be measured across the entire range from 10^{-6} to 10^{-10} g/cm^2—each works only for a given part of the range, which is why the methods are so numerous. This poses the important problem of creating methods for quick measurement of surface cleanliness that would use relatively simple equipment, offer short measurement times, prevent any mechanical damage to the substrate, preserve its properties, and not involve the use of special probes with surfaces that require cleaning.

Methods for final cleaning and surface treatment used to increase the adhesion of thin metal films being deposited on the surface and methods for plasma etching are based on the use of low-temperature plasma in the form of a large-format flux with uniform particle distribution over its cross section. Particles in the flux must move along the normal to the surface being treated—that is, the movement must be anisotropic. Currently, plasma etching and final cleaning are widely used for both modern optics [1–15,24–33] and micro- and nanoelectronics [34,35]. The active component in these techniques is low-temperature plasma formed by glow discharge, HF discharge, SHF discharge, and magnetron discharge [35,36].

To generate large-format plasma fluxes with glow discharge, hollow-cathode, hollow-anode sources are used [37–42]. But this approach poses the problem of suppressing the effect of discharge instability on discharge uniformity. References 37 and 38 solve this problem by using microhollow-cathode discharges as plasma cathode. This makes it possible to create stable glow discharges having large volume and surface area. Plasma uniformity is achieved by using magnetic-field systems or systems combining magnetic and electrostatic confinement of fast electrons in a large-aperture hollow cathode and generating ion-emitting plasma in the anode cavity [39,40].

Creating a uniform large-format plasma flux in HF, SHF, and magnetron discharge sources is yet another challenge [43–46]. To cite one example, with SHF gas discharge as the source, as the gas pressure in the working chamber drops, the parameters of the formed plasma are increasingly affected by electron–cyclotron resonance due to the nonuniform magnetic field of the solenoid [47]. This, in turn, results in nonuniform treatment of large-diameter wafers. The authors of References 48–51 propose increasing plasma uniformity by changing the shape and design parameters of HF and SHF discharge sources. It has been noted that the radial localization of the section with the highest plasma density tends to deviate from the center as the plasma chamber's height decreases and pressure increases: the vacuum-chamber area where the discharge is excited affects plasma uniformity. The influence of these factors on plasma uniformity can be minimized by broadening the plasma chamber; using specially shaped guiding chambers [46]; using two coils—planar and vertical—simultaneously to excite a high-frequency field [45]; and using specialized annular antennas and magnetic systems [43,44].

In plasma-chemical etchers that have been used thus far, gas discharge generates plasma in the electrode gap. With such etchers, increased uniformity requires more complex and energy-intensive plasma sources, and the following detrimental factors typical of vacuum plasma systems and plasma-forming discharges are not eliminated:

- An increase in the relative surface area being treated leads to a decrease in the etch rate [52,53], including through ion shadowing, neutral shadowing, and the microloading effect [54].
- System parameters (high-frequency offset, power, pressure, etc.) need to be optimized [54].
- The collision of plasma with reactor walls results in the formation of an additional polymer source and a higher tendency for etching to stop [54].
- The material, shape, and properties of the substrate surface affect the parameters of gas discharge [55,56].
- The surface being treated becomes contaminated with low-active or inactive plasma particles [57–59], resulting in changes in etching characteristics.
- The polymerization effect occurs during etching in fluorocarbon gases [60,61].
- Charged particles' parameters depend on the operating regimes of gas-discharge devices.

These factors increase the complexity of final-cleaning and etching processes and make it difficult to determine the parameters of optimal regimes and to form large-format plasma fluxes that enable uniform cleaning and etching across the substrate surface regardless of its size. These drawbacks result in higher product costs and indicate a need for gas-discharge devices to form plasma fluxes.

When low-temperature plasma impinges on the surface of the material being treated, the etching area must receive only negatively charged fluorocarbon-gas particles. Fluorocarbon gases improve the substrate's etching anisotropy and rule out the possibility of discharge products being accumulated at the surface [62]. Plasma (plasma fluxes) must be directed and generated outside the electrodes of the gas-discharge device. Charged and chemically active particles in plasma should not collide with the working chamber's sidewalls (plasma should be localized); the charged particles' parameters should not depend on the gas-discharge device's operating regimes; and the particles should be uniformly distributed over the flux cross section. As may be concluded from several publications [63–66], these properties are possessed by ion–electron beams generated by high-voltage gas discharge outside the electrode gap, as well as by high-voltage gas-discharge off-electrode plasma, a type of plasma that was first studied and tested at the Korolev State Aerospace University, Samara, Russia.

High-voltage gas discharge was first discovered in the 1970s at the Paton Electric Welding Institute of the Academy of Sciences, the Ukrainian Soviet Socialist Republic, and was successfully used in welding and laser technologies [67–69]. In the 1980s, the authors of References 64 and 65 expanded the area of its application: high-voltage gas discharge was used for soldering semiconductor components. The authors of this monograph have created several plasma-chemical etching and ion-chemical etching reactors that generate directed low-temperature plasma fluxes through a high-voltage gas discharge outside the electrode gap. We will refer to this type of plasma as *off-electrode plasma*.

Scientists at the Korolev State Aerospace University have developed a device for quick nondestructive evaluation of surface properties that is free from the shortcomings typical of similar devices made in Russia and abroad [70,71]. Given that the reactors and the device are fundamentally new, we find it interesting to study the possibility of using them to fabricate micro- and nanoreliefs and diffraction microstructures on large-format wafers in low-temperature off-electrode plasma. This book is our first attempt to achieve this goal—one that opens new horizons for furthering the study of gas discharge physics, plasma physics, solid-state physics, and diffractive and computer diffractive optics.

The preparation of this book has been supported by the Ministry of Education and Science of the Russian Federation as part of the analytical departmental special-purpose program titled Development of the Scientific Potential of the Higher School for 2006–2008; grants NS-1007.2003.01, NS-3086.2008.9, MK-3038.2007.9, and MD-5205.2016.9 from the Russian president in support of young Russian scientists and leading scientific schools; grant from the Samara Oblast government and the US Civilian Research and Development Foundation (CRDF project RUX0-014-SA-06) as part of the Russian–American Basic Research and Higher Education Programme; and grants from the Russian Foundation for Basic Research (projects 07-07-97601, 04-02-08094, and 16-07-00494 A); and the 2008–2009 program titled Participant of Youth Science and Innovation Contest, run by the Foundation for Assistance to Small Innovative Enterprises in Science and Technology (state contract 6638P/8718).

We wish to express our sincere appreciation to V. A. Soifer, a corresponding member of the Russian Academy of Sciences, for his active support over many years and for making it possible to advance the study of this scientific field. Our heartfelt thanks go to A. A. Orlikovskiy, a full member of the Russian Academy of Sciences; M. Ye. Sarychev, a doctor of physics and mathematics; M. G. Putrya, a professor and a doctor of technical sciences; V. A. Bykov, a doctor of technical sciences; A. A. Golishnikov, a candidate of technical sciences and an associate professor; V. V. Losev, a candidate of technical sciences; V. I. Chepurnov, a candidate of technical sciences and an associate professor; the team of Prospective Technologies and Equipment for Micro- and Nanoelectronics (a regularly held scientific workshop at the Institute of Physics and Technology, the Russian Academy of Sciences); and the team at the Image Processing Systems Institute (the Russian Academy of Sciences), for many useful, fruitful, and efficient discussions, as well as invaluable help with recording precise measurements, without which this book would not have been possible.

SUMMARY

This book is the outcome of research into a new scientific topic: microstructuring the surfaces of optical materials with directed fluxes of off-electrode plasma generated by high-voltage gas discharge and developing the methods and equipment related to this technique. The book describes the results of theoretical and experimental studies on the electrical and physical properties of high-voltage gas discharges used to generate plasma outside an electrode gap. The authors draw on their complex experimental research to offer physico-mathematical models that illustrate the interaction of off-electrode plasma particles with atom–molecule complexes at the surface of a solid, as well as with heterostructures and polymers.

The book also discusses a new class of methods and devices that makes it possible to implement a series of processes for fabricating diffraction microstructures on large-format wafers by using low-temperature plasma and to broaden the existing range of diffractive optical elements in order to meet the prospective requirements of computer diffractive optics as the field develops.

This book will be useful to scientists and specialists from a broad range of fields such as gas discharge physics, plasma physics, solid-state physics, micro- and nano electronics, and diffractive and computer diffractive optics.

Authors

Nikolay L. Kazanskiy is a Doctor of Science, and head and acting director of the Diffractive Optics Laboratory at the Image Processing Systems Institute and a professor in the Technical Cybernetics Department at the Samara University, Samara, Russia. He is a member of SPIE and IAPR, the author and coauthor of 240 articles and 10 monographs, and a coinventor of 46 patents in diffractive optics, mathematical modeling, and nanophotonics.

Vsevolod A. Kolpakov is a doctor of physics and mathematics (Doctor of Science) and a professor in the Department of Electronic Engineering and Technology at the Samara University, Samara, Russia. He is an expert in ion-plasma technology and quality management, the author and coauthor of 120 scientific publications, including 3 monographs, 2 textbooks, and 40 articles, and a coinventor of 9 patents.

1 Forming Directed Fluxes of Low-Temperature Plasma with High-Voltage Gas Discharge outside the Electrode Gap

Modern diffractive optics interrelates with the rapidly advancing studies of ion-plasma processes. For purposes of this monograph, we will divide the studies into two groups: studies of physical processes occurring in plasma and studies of interaction between plasma particles and the surface being treated.

Because these topics have been covered in several reviews [34,35,72–83], this chapter discusses only those points that are necessary for fully understanding the nuances inherent in forming directed fluxes of low-temperature plasma generated by a high-voltage gas discharge outside the electrode gap.

1.1 OVERVIEW OF DEVICES USED FOR GENERATING LOW-TEMPERATURE HIGH-VOLTAGE GAS-DISCHARGE PLASMA

A characteristic feature of fabricating diffractive optical elements (DOEs) is the formation of a diffraction microrelief across large areas with trenches having almost identical parameters. For this purpose, ion-plasma technology for fabricating DOEs requires the use of plasma particles as a process tool whose energy varies from one to a few to hundreds of electronvolts and whose concentration and energy are uniformly distributed over the cross section of the plasma flux. The simplest way to achieve this condition is to use axially symmetric beams (fluxes) of low-temperature plasma. Currently, high-voltage gas-discharge devices are used for generating such beams [63]. As noted in the preface, these beams are widely used in technology, and so the physical processes occurring in systems generating the beams are of particular interest.

High-voltage discharge in gas is an abnormal type of glow discharge: it has all the advantages of glow discharge and does not have the disadvantage of gas-discharge parameters' being dependent on the location and surface properties of the substrate. The occurrence of a high-voltage discharge is caused by nonuniform distribution of the electric field, achieved through the use of specially designed electrodes in gas-discharge devices (see Figure 1.1) [63].

Analysis of the electrode design shows that if electrodes are made of a solid material and the anode and cathode are brought together to within the Aston dark space,

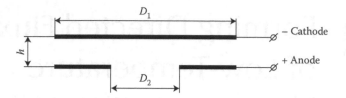

FIGURE 1.1   Schematic of a gas-discharge device based on a unit cell generating arbitrarily shaped ion–electron beams in high-voltage gas discharge.

glow discharge disappears because the inequality $\gamma Q < 1$ is satisfied, where γ is the number of electrons knocked out from the cathode by one Q ion. But if a through-hole is made in the anode, satisfaction of the inequality $\gamma Q \geq 1$ will not be inhibited in its area. Physically, this means that one or more electrons take part in generating one or several pairs of positive Q ions, thus providing conditions for a gas discharge outside the anode.

Therefore, in the anode-hole area, the electron knocked out from the cathode surface under the action of a field gradient continues moving in the accelerating field of the gas-discharge device outside the electrode gap until residual-gas atoms are ionized. As a result, a directed flux of electrons is generated in the anode-hole area. Simultaneously, every point of the cathode has its own electron-movement path along which, on average, Q ions are generated that reach the cathode at the point of electron escape. A discharge can appear only when the electrons and ions move along the same path [63].

Analysis of equipotential lines [65,84] reveals that this condition is best satis-fied near the symmetry axis of the anode hole, while toward the periphery of the anode hole, this condition is least satisfied owing to deformation of the field lines (see Figure 1.2).

Thus, the mechanism of discharge formation eliminates the conditions for anode sputtering.

If the electric-field gradient in the cathode is sufficient, every electron knocked out from it moves along the direction corresponding to the straight section of the relevant field line [63]. Because of their low mass, electrons gain significant speed along the mean free path and, when a field line loses straightness, they run off the line and jump onto the straight section of an adjacent field line.

But typically, ions that have large masses and low speeds move strictly along the direction of the field line. Therefore, in the area where the electron-movement path bends, electrons and ions cease interaction, thereby violating the conditions required for a self-sustaining gas discharge.

The number of ions generated in a straight near-cathode field-line section depends on the section's length, gas pressure, and distribution of electric-field lines along that section. In particular, the ion concentration increases with an increase in pres-sure, but not all ions generated in the straight section of the field line will move toward the cathode along the field line: only those will do whose movement starts in places where the field gradient is sufficient to ensure that they move toward the cathode.

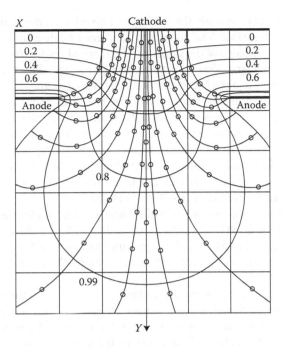

FIGURE 1.2 Electric-field distribution in the electrode area of the gas-discharge device.

Therefore, the discharge occurs only in straight sections of the field lines, provided that electrons and ions in such a section succeed in gaining an amount of energy equal to or higher than that required for ionization of residual-gas atoms and the cathode material.

References 63, 67, and 85 describe devices used for generating ion–electron beams in a high-voltage gas discharge. Such devices incorporate an electrode system that consists of a flat cathode, flat anode, and diaphragm with an aperture whose shape and size d_a may vary. The distance h between the cathode and the diaphragm ensures that, at an operating gas pressure of $p = 10^{-2}$–10^{-1} Torr and a voltage of ≤ 50 kV, discharge cannot occur outside the diaphragm aperture. The distance h in the studies varied from 5 to 15 mm and the size d_a from 10 to 20 mm. An ion–electron beam current of 45–600 mA is achievable at an electrode voltage of 17–25 kV. A gas-discharge device operating under such conditions cannot be used for fabricating diffraction microreliefs.

Reference 64 describes the possibility of using a high-voltage gas discharge to generate low-temperature plasma beams with ribbon-shaped cross sections. In that study, a 20-mm-thick anode was used with a hole in the middle sized $d_a = 5 \times 30$ mm. The distance h was 3 mm to stabilize the high-voltage discharge and lower its ignition potential. The beam current controlled by varying the accelerating voltage over the 0.5–7 kV range was 0–40 mA. A gas-discharge device operating under these conditions can be used for etching materials [86]. The energy of the particles colliding with the substrate surface varies from a few to hundreds of electronvolts and is sufficient for plasma-chemical or ion-chemical etching. But the beam's geometric dimensions in this case are only 0.8×25 mm. This necessitates scanning the beam across the substrate surface. The authors of Reference 64 determined that when the ribbon beam nears the edge of the substrate, the thermal field starts to be intensively reflected from its end face. This leads to an intensive accumulation of heat energy in the surface area between the beam and the substrate edge, thus causing a temperature rise. Different etching conditions prevail on the edges and in the central part of the wafer. This difference does not allow a diffraction microrelief to be formed with uniform parameters over the entire substrate surface. To eliminate this drawback, in most practical cases, the total substrate area is significantly increased and conditions are created for uniform heating in the etching area. Significant consumption of costly substrate material and a low etch rate have prevented widespread use of ribbon beams for forming diffraction microreliefs in DOEs.

This monograph proposes a new gas-discharge device for forming optical microreliefs. The device is capable of creating directed large-format fluxes of low-temperature plasma generated by a high-voltage gas discharge outside the electrode gap (see Figure 1.3), with active plasma particles moving along the normal to the substrate surface.

To increase electron emission, the cathode is made of aluminum [87], and to ensure the plasma particles' energy is distributed more uniformly, the anode is made of a 1.8×1.8 mm mesh of 0.5 mm stainless-steel wire, which significantly decreases the anode's chemical interaction with plasma particles and improves its resistivity to thermal heating [88]. The inter-electrode distance is selected according to the inequality $h \leq 2\text{–}3\lambda$, where λ is the mean free path of a charged particle. This makes it possible to decrease the ignition voltage to $U_i = 300$ V. In this case, the discharge current varies from 0 to 200 mA and the accelerating voltage, from 0.3 to 6 kV. The electrode diameter is selected according to the dimensions of the wafer being treated, and it may vary from 50 to >200 mm. The cell size of the mesh has been selected with the following conditions in mind: If the cell size is greater than 1.8 mm, the uniformity of charged-particle distribution over the cross section of the

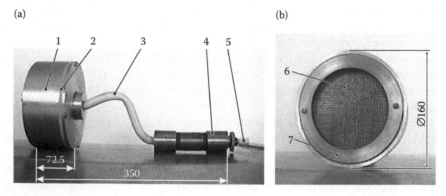

(a)   (b)

FIGURE 1.3 Lateral (a) and frontal (b) view of the high-voltage gas-discharge device: (1) housing-anode, (2) cathode-housing insulation, (3) high-voltage cable gland insulation, (4) coupling for attaching gas-discharge device to working chamber, (5) high-voltage cable gland, (6) anode gauze, and (7) retaining ring of anode gauze. Overall dimensions are in millimeters.

plasma flux decreases because of the presence of fewer axial field lines, resulting from the presence of fewer cells. And in the case of $d_a < 1.8$ mm, the conditions for self-sustaining discharge are violated within the specified ranges of its currents and accelerating voltages at $h \leq 2$–3λ, and this does not contradict the results presented in Reference 63.

The formation and existence mechanisms of high-voltage gas discharges should be studied with a view to efficiently using gas-discharge–generated low-temperature plasma in ion-plasma etching of DOE microreliefs, to enhancing the functionality of the gas-discharge device, and to using it as a basis for designing new equipment to form directed plasma fluxes.

1.2 FEATURES OF LOW-TEMPERATURE OFF-ELECTRODE PLASMA GENERATED BY HIGH-VOLTAGE GAS DISCHARGE

For purposes of this study, a UVN-2M-1 vacuum evaporator was used with a working chamber containing a gas-discharge device (see Figure 1.3a and b) for generating off-electrode plasma. Halocarbon-14 (CF_4) was used as process (working) gas.

This arrangement leads to appreciable deformation of the electric-field lines near the anode hole (see Figure 1.2) [63]. A feature of the electric-field distribution is the increased length of the straight section of the field line toward the symmetry axis of the anode hole. In the area of the hole edge, the length of the straight section is less than the length of the electron's mean free path, and no high-voltage discharge occurs.

As they move along straight sections of field lines, the electrons emitted by the cathode under the action of the field gradient gain sufficient energy to result in ionization of the residual-gas atoms outside the electrode gap. Most positive ions appear on straight sections of field lines in the axial area of the anode hole and reach the cathode surface at points where electrons are emitted; this is confirmed by geometric parameters of the spots created on the cathode surface by the positive ions (see Figure 1.7). The spots are shaped similar to the mesh cells, but their dimensions are half as little—this allows us to regard them as the dimensions of the axial area participating in the self-sustaining of the discharge.

Plasma parameters were measured through the collector method [89], rotating-probe method [87], and method for determining flow parameters of charged particles [90]. To exclude sputtering, the probe was fabricated from a 0.1-mm tungsten wire, thus practically eliminating any impact on the plasma parameters.

The reactor's working chamber was evacuated to an initial pressure of 10^{-4}–10^{-2} Torr. By introducing the process gas, a pressure of 10^{-1}–10^{-2} Torr was set. Then, the cathode was energized by a BP-150 power supply unit supplying a voltage equal to or higher than the ignition voltage. Under these conditions, outside the electrodes in the area of each cell of the gauze anode, a high-voltage gas discharge is ignited that generates a plasma flux directed toward the substrate surface. Thus, the plasma generated by the gas-discharge device is the aggregate of plasma fluxes formed by each cell of the gauze anode. Analysis of the gas-discharge device described in Reference 63 supports the above statement: the analysis shows that each cell of the anode mesh represents a hole and that the entire flux of charged particles is

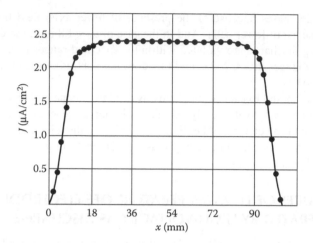

FIGURE 1.4   Distribution of charged particles across the plasma flux.

composed of identical microfluxes. The microflux parameters depend on the cell size and cathode-surface properties, which are identical in the case under study, as are the parameters of the individual microfluxes. As a result, charged particles are uniformly distributed over the flux cross section, and nonuniformity results only from the edge effect of cathode design. The edge-effect area is minimal: it does not exceed 12 mm (see Figure 1.4).

For the parameters under study, we used the equation $J = f(X)$ (where J equals I_z/S; I_z is the beam current measured with a moving collector with a 0.8-mm hole [89], and S is the hole area; see Figure 1.4) to evaluate the uniformity of charged-particle distribution. This equation shows that the uniformity over the cross section of the high-voltage gas-discharge plasma flux was not worse than 98% in the range 12 mm $< X <$ 90 mm. These results correspond to those found in Reference 91.

The existence of outside-electrode discharge suggests that the discharge particles are in free motion [63]. This sharply reduces the impact of the device's operating regimes on particle parameters, thus practically eliminating the loading effect and the necessity of protecting the cathode from sputtering. Free motion of the particles and sharp discharge boundaries suggest that, outside the anode, the particles move straight and perpendicularly to its surface.

Actually, Figure 1.4 shows that the distribution of the charged particles across the plasma flux is uniform and that its motion toward the sample surface is perpendicular.

Analysis of the V–I curve of the discharge (see Figure 1.5) indicates that its formation depends on ionization of process-gas atoms (α process) and the cathode material (γ process) [92]. Note that in the voltage range of $300 \leq U \leq 1,000$ V, ionization of process-gas atoms is predominant, while at $U \geq 1,000$ V, intense cathode sputtering takes place, leading to ion–electron emission being responsible for the remaining section of the V–I curve.

But in the region of relatively low pressure ($p \leq 1.5 \cdot 10^{-1}$ Torr) in the range $20 \leq I \leq 50$ mA, there is a pronounced section in the V–I curve where I-dependence is weak. This suggests that at high pressures in this voltage range, the electrons

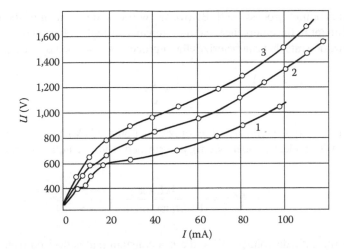

FIGURE 1.5 $V–I$ curve of high-voltage gas discharge at various pressures in the working chamber: (1) $1.5 \cdot 10^{-1}$ Torr, (2) $1.2 \cdot 10^{-1}$ Torr, and (3) $9 \cdot 10^{-2}$ Torr.

manage to gain sufficient energies for ionizing process-gas atoms, thus actively contributing to an increase in current with even a small increase in voltage. This assumption agrees with the graph in Figure 1.6 [93]: voltage saturation in the pressure range $1.8 \cdot 10^{-1}$ Torr $\geq p \geq 9 \cdot 10^{-2}$ Torr in the case of a clean (new) cathode proves that the capabilities of process-gas ionization have been exhausted, and sputtering and ionization of the cathode atoms (ion–electron emission) are responsible for the rise in the curve at $p < 9 \cdot 10^{-2}$ Torr.

To prove the above statements, we will evaluate the parameters of gas-discharge existence. It is known that the ionization of process-gas atoms can result from electron (α process) and positive-ion (β process) action and that electron emission from the cathode surface can be caused by ion bombardment (γ process) and radiation-induced

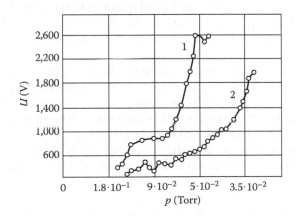

FIGURE 1.6 Discharge voltage versus pressure: (1) clean (new) cathode and (2) contaminated cathode (after a long period of operation).

surface ionization (δ process) [92]. Below, we will elucidate which of the processes are predominant in the emergence and maintenance of high-voltage gas discharge.

The ionization rate that characterizes the α process is described by the equation [94]

$$\alpha_i = \frac{1}{l_i} = \frac{E}{\phi_i}, \tag{1.1}$$

where l_i is the ion range (cm), ϕ_i is the ionization potential (V), and E is the intensity of the nonuniform electric field (V/cm), determined from the equation [65]

$$E(y) = \frac{4cU / \pi}{(1+h/c\pi) \, 4c^2 + y^2}, \tag{1.2}$$

where U is the cathode voltage (V) and c is a constant that is derived from a system of equations [65] and that equals 0.08 cm for a 1.8 × 1.8 mm anode hole. To calculate the strength of the electric field acting on a charged particle at the first length of its mean free path, λ (cm), we must replace the y in Equation 1.2 with the value of λ calculated as follows:

$$\lambda = \frac{4\sqrt{2}}{n_0 \sigma}, \tag{1.3}$$

where n_0 is the concentration of halocarbon-14 molecules, and it equals $0.29 \cdot 10^{16}$ cm^{-3} for a pressure of $9 \cdot 10^{-2}$ Torr, and σ is the effective cross section, equal to $14.1 \cdot 10^{-16}$ cm^2 [94,95]. As the discharge was ignited at a minimum accelerating voltage of $U = 300$ V at the cathode, by substituting into Equation 1.2, U, the known h and c, and λ, which equals 1.3 cm according to Equation 1.3, we obtain $E = 17.3$ V/cm. Substituting this value of electric-field strength into Equation 1.1 yields $\alpha_i \approx 1$ cm^{-1}, which corresponds to the condition for outside-anode gas discharge ($\gamma Q \geq 1$). Comparing the values of λ and l_i at the above electrode voltage yields $\lambda > l_i$, thus suggesting that residual-gas molecules can be ionized [95].

The efficiency of the positive ion-induced ionization of process-gas molecules is low, and therefore the β process is dispensable in gas-discharge studies [94]. Because high-voltage discharge is independent, with no extra irradiation sources found in the vacuum chamber, the δ process, too, is dispensable. Therefore, positive ions are the major source of cathode-emitted electrons. The contribution of positive ions to the knocking out of electrons from the cathode surface is characterized by the secondary-emission coefficient, which equals $\gamma_e = 7.16 \cdot 10^{-5}$ if calculated as described in Reference 96 for $U = 300$ V. Given a cathode voltage of 1,000 V, the above-discussed calculation techniques [94,96] give the following values of the coefficients α_i and γ_e: $\alpha_i \approx 4.8$ cm^{-1} and $\gamma_e = 0.66$. Comparing the two values, we can see that there is an almost five-fold increase in the volume ionization of process-gas molecules, while ionization due to ion–electron emission has increased by a factor of 10^4. Thus, for cathode voltages of $300 \leq U \leq 1,000$ V, process-gas ionization is mainly caused by volume ionization owing to electron impact. For $U \geq 1,000$ V, the major ionization mechanism is ion–electron emission, and this agrees well with the graphs in Figures 1.5 and 1.6.

Violation of the exponential dependence in Figure 1.6 in the range $p = 5.5 \cdot 10^{-2}$–$4.8 \cdot 10^{-2}$ Torr is attributable to the emergence of unstable microarch discharges between the cathode and anode, which can be seen with the naked eye. The conditions for the emergence of this type of parasitic discharge in these voltage and pressure ranges are similar to those for high-voltage discharge. Therefore, the two emerge virtually simultaneously. As the pressure continues to decrease and the ignition potential increases, one of the discharges starts to prevail, and a breakdown of the dielectric space between the anode and cathode ensues. Traces of three such breakdowns are shown in Figure 1.7.

The absence of saturation in the case of the contaminated cathode (Figure 1.6) (after a long period of operation) suggests that there are structural changes on the cathode surface, as can be seen in Figure 1.7. These appear during operation under the action of plasma-flux microrays as traces of sputtering, reproducing anode-hole contours in the process [97].

It is known [98] that any disturbance of the crystalline lattice causes weakening of inter-atomic bonds. Because of ion bombardment, such disturbances possess a low ionization potential as compared with the core material. Thus, the potential of the high-voltage discharge ignition should be expected to decrease in accordance with the form of the curve in Figure 1.6. In this case, the character of the curve is determined by the predominant emission of the cathode material, which begins at a lower pressure. Low pressure facilitates eliminating from the cathode surface easily evaporated contamination resulting from adsorption of various atoms and molecules, and this leaves ion–electron emission as the only mechanism for sustaining discharge.

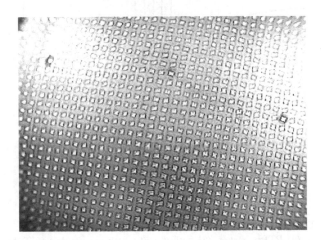

FIGURE 1.7 Breakdown traces and 0.9×0.9 mm spots formed on the cathode surface by positive ions (view of the cathode surface after a long period of operation).

1.3 DESIGN CHANGES TO THE HIGH-VOLTAGE GAS-DISCHARGE DEVICE

In terms of practical use in the technology for fabricating optical microreliefs described in Sections 1.1 and 1.2, it would be most efficient to use the high-voltage gas-discharge device shown in Figure 1.8 [58,59].

But the device's operation has shown that, at the operating voltage in the system comprising cathode, cathode insulator, fluoroplastic (PTFE) gasket, anode, and mesh, conditions arise for the occurrence of a surface breakdown (see pos. 2 in Figure 1.9a) that changes uncontrollably to an arc discharge (see pos. 1 in Figure 1.9a and b). This results in the cathode insulation getting burnt (see pos. 4 and 6 in Figure 1.8) and in a carbon film forming on the breakdown surface, thereby leading

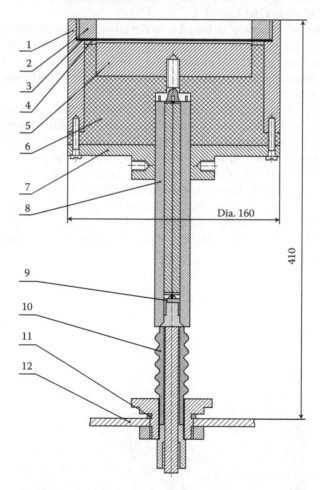

FIGURE 1.8 Design of the high-voltage gas-discharge device: (1) housing, (2) anode ring, (3) gauze anode, (4) PTFE gasket, (5) cathode, (6) cathode insulator, (7) housing lid, (8) high-voltage cable, (9) contact cap, (10) high-voltage cable gland, (11) rubber gasket, and (12) working-chamber bottom. Overall dimensions are in millimeters.

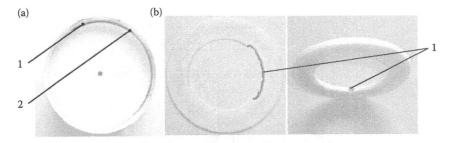

FIGURE 1.9   (a and b) External view of the cathode insulator and PTFE gasket after electric breakdown: (1) traces of arc discharge-induced breakdown and (2) traces of surface discharge–induced breakdown.

to a short circuit in the anode–cathode system and to permanent failure of the device (see pos. 1 in Figure 1.9a and b).

In addition, at high voltages of $U = 3$–5 kV, a glow discharge occurs between the surface of the power cable and those of the working chamber's wall and process tooling, resulting in the insulation of the cable and the high-voltage cable gland getting fully burnt. This makes the device much more difficult to operate and, because the insulating materials are expensive, more costly. To remedy these shortcomings, this section proposes modifying the design of the cathode insulation, gauze anode, power cable, and high-voltage cable gland.

The dielectric properties of cathode insulation have been improved by making an n-shaped cut in the cathode insulator, filling it with vacuum oil, and sealing it with a ring-shaped PTFE gasket (an insert) and gaskets made of IRP-1015 rubber for vacuum applications (see Figure 1.10).

When the insulator is attached to the insert, the vacuum oil overflowing from the n-shaped cut oozes out during sealing, thus forming close-to-monomolecular layers at the side surfaces of the insulator and insert. These layers fill all micro- and macro-pores at the surfaces of the contacting materials, thereby preventing the occurrence of parasitic microdischarge. This can be proved by the following observation: With the modified high-voltage gas-discharge device installed in the working chamber of the vacuum system, excess vacuum oil continues to be drawn out under the action of the pressure gradient during the first evacuation stage and reaches the cathode's peripheral surface in the area of the PTFE gasket's counterbore. After the working chamber is depressurized, the oil can be removed with alkaline solutions.

Thus, in the case of a breakdown, free carriers move through vacuum oil layers with high dielectric properties. Even if a breakdown occurs, the oil fills the breakdown spot, thus restoring the insulation capability of the cathode insulator.

The total dielectric thickness between the cathode and the housing, L_1, equals 5 mm, and it depends on the breakdown voltage of PTFE, which is equal to 26 kV/mm [99]. The PTFE gasket's thickness d_4 is determined by the anode–cathode distance, which, in turn, is determined by the maximum operating pressure in the vacuum chamber and should be less than the Aston (cathode) dark space of a glow discharge. Otherwise, some type of glow discharge will ignite between the electrodes.

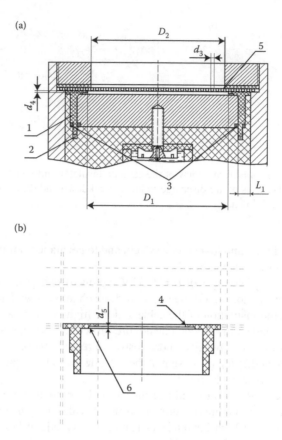

FIGURE 1.10 Design of cathode insulation (a) and PTFE gasket (b): (1) n-shaped cut, (2) VM1 vacuum oil, (3) vacuum rubber, (4) PTFE gasket end, (5) copper ring, (6) PTFE gasket counterbore, (D_1) inner diameter of PTFE gasket, (D_2) inner diameter of anode ring, (d_3) gauze-anode wire diameter, (d_4) minimum thickness of PTFE gasket, and (d_5) anti-break-down gap thickness.

An electric breakdown of the cathode–anode assembly depends on the length of the dielectric surface along the perimeter of the n-shaped cut and the PTFE gasket. The length should be sufficient to prevent parasitic discharge on the surface. A breakdown on the upper inner edge of the PTFE gasket occurs owing to cathode sputtering and formation of a carbon film (organic impurities) on the cathode. Aluminum and carbon atoms adsorbed on the PTFE gasket's ends gradually form a conducting layer, thereby leading to anode–cathode short circuit, occurrence of parasitic discharge owing to power generation, and breakdown of the PTFE gasket. To prevent such a breakdown, a counterbore with a diameter of 125 mm and a height of $d_5 = 1$ mm should be cut in the inner surface of the PTFE gasket (see Figure 1.10b). This significantly increases the end surface of the gasket and rules out the possibility of aluminum atoms and contaminants depositing on the inner surface of the counterbore. For the same purpose, diameter D_2 of the anode ring (110 mm) is selected to be less than the inner diameter D_1 of the PTFE gasket, which is equal to 123 mm.

Thus, the cathode–anode insulation comprises the cathode insulator with its n-shaped cut partially filled with vacuum oil and the PTFE gasket whose counterbore serves as an anti-breakdown gap that not only protects the insert from contamination but also lengthens the path of charged particles in the case of a surface breakdown [94].

The modified cathode insulation helps prevent the mechanisms of surface breakdown over the entire range of operating voltages at the electrodes.

To rule out the possibility of a short circuit in the cathode–anode system as a result of the anode mesh sagging because of heating during the device's operation, the gauze anode should be modified by welding it to a copper ring having the same inner and outer diameters as those of the anode ring (see Figures 1.8 and 1.10). Before welding the mesh to the ring, the mesh–ring system must be cooled to a temperature from −20°C to −30°C. The difference between the linear expansion coefficients of the materials in contact will tighten the mesh and prevent it sagging when the mesh–ring system heats up during the operation of the gas-discharge device.

Gas-discharge devices are supplied with electric power through high-voltage cables such as 3KVEL, 4KVEL, and 3KVR-75 [99]. Immediate contact between the conductor and the insulation results in a charge being accumulated on the outer surface of the insulation, and parasitic discharge occurs between the insulation and the surfaces of the working chamber and process tooling. Preventing such a discharge requires an excessive increase in dielectric thickness, as well as use of more complex cables with multilayer insulation made of expensive materials. To overcome these shortcomings, we propose using a power cable made of affordable materials that can be easily bent to obtain desired shapes and eliminate the conditions conducive to occurrence of parasitic discharge. This high-voltage cable incorporates a PTFE sheath in the form of a hollow cylinder (see Figure 1.11) with an outer diameter of 16 mm and a wall thickness of about 2 mm [100,101]. Precise wall thickness depends on whether the cable needs to be bent when the gas-discharge device is being installed into the vacuum system: the sharper the bend, the thinner should the sheath wall be. The sheath contains a copper wire with stabilizing PTFE disks loose-fitted on it, alternating with ceramic or PTFE tube-shaped separators that allow the wire to be firmly positioned centrally inside the sheath. This ensures uniform distribution of the electric field along the entire length of the cable.

Disks and tubes serve to prevent the wire touching the sheath surface, thereby increasing the insulation-breakdown voltage while allowing the cable to be bent. The outer diameter of a disk is $0.9D$, where D is the inner diameter of the sheath. The disks have a rounded outer surface and a cross section in the form of a $30° < \chi < 60°$ angle.

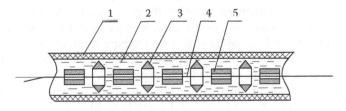

FIGURE 1.11 Design of high-voltage cable—a segment: (1) PTFE sheath, (2) vacuum oil, (3) stabilizing PTFE disk, (4) copper wire, and (5) PTFE separator.

This ensures that along the generating line, there is minimum contact between the sheath's inner surface and the disk. This geometry provides a point contact between the disks and the sheath's inner surface when the cable is bent. If the angle χ is selected to be less than 30°, bending the cable crushes the disks, irreversibly changes their geometry, changes the sheath's inner dimensions, and affects the cable's dielectric properties because of changes in the electric-field distribution. If χ values exceed 60°, cable flexibility notably deteriorates because the sheath's inner surface touches the PTFE disks' outer surfaces when the cable is bent sharply. Bending the cable further stretches the sheath material, irreversibly deforms it, and affects its dielectric properties.

The separating tubes have a length of $0.5–1D$ and an outer diameter of $0.2–0.3D$. The tubes serve to keep the stabilizing disks spaced at a fixed distance of $0.5–1D$. A distance less than $0.5D$ affects the cable's flexibility and is undesirable for cables used in the vacuum system's working chamber. A distance exceeding D deforms the sheath's shape, thereby changing the distance between the cable's conductor and the sheath's inner surface. If the stabilizing disks' and tubes' outer diameters are less than the sheath's inner diameter, the conductor can be easily moved inside the sheath; an oil film forms between the surfaces of disks, tubes, and the sheath's inner surface; deformation of the sheath can be closely monitored when the cable is being bent; and the copper wire can be reliably isolated from the working chamber's environment.

VM-1 oil is used as a liquid insulator because with its good dielectric properties and a low vapor pressure, the oil is well suited for vacuum applications.

After the PTFE sheath is filled with oil, the cable is connected to the high-voltage cable gland, and the unit is installed into the working chamber of the vacuum system.

Given that the oil-filled high-voltage cable needs to be connected to the vacuum system through a ceramic insulator, we designed an appropriate high-voltage cable gland (see Figure 1.12).

This high-voltage cable gland consists of a ceramic insulator and shielding elements comprising a housing and upper and lower housing lids. The cavity of the high-voltage cable gland is filled with vacuum oil as well. This increases the insulation-breakdown voltage and rules out the possibility of parasitic discharge between the ceramic insulator and the housing. Rubber seals, a washer, and a retaining nut serve to seal the assembly. A PTFE sleeve makes the cable more rigid when the compressive force begins to act on it from the top rubber seal.

Long-term operation of the high-voltage cable gland has shown that even when the gas-discharge device is operating at full capacity ($I = 200$ mA, $U = 6$ kV, $p = 10^{-1}$ Torr) for a long period (exceeding 100 h), no parasitic discharge occurs.

Figure 1.13 shows the improved design of the gas-discharge device. The assembly consists of a housing, anode ring, anode mesh with a copper ring, cathode insulator with an n-shaped cut, PTFE gasket with a counterbore forming an anti-breakdown gap, cathode, rubber gasket, PTFE insert, housing lid, modified high-voltage cable, modified high-voltage cable gland, and retaining screws.

The cathode is made of aluminum in the form of a 20-mm-thick disk having a diameter of 128 mm, machined to a finish grade of $\Delta12–\Delta14$. Aluminum yields maximum electron emission because its work function is 4.25 eV. To prevent concentration of electric-field intensity, the cathode edges are rounded ($R = 1–2$ mm, where

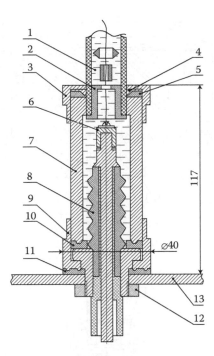

FIGURE 1.12 High-voltage cable-gland assembly and a segment of high-voltage cable: (1) high-voltage cable, (2) PTFE sleeve, (3) upper housing lid, (4) washer, (5) upper rubber gasket, (6) contact cap, (7) shielding housing, (8) ceramic insulator, (9) lower housing lid, (10)–(11) rubber gaskets, (12) retaining nut, and (13) working-chamber bottom. Overall dimensions are in millimeters.

R is the rounding radius). The cathode insulator is a cylindrical PTFE part with a cavity for the cathode and the n-shaped cut for the PTFE gasket.

The cavity houses the cathode, which is screwed to the insulator via a rubber gasket. The screw serves as both a fastener and an electrical contact for the high-voltage cable. The n-shaped cut is filled with vacuum oil such as VM-1 and the PTFE gasket is loose-fitted into the cut. As noted above, the cut and the gasket lengthen the area where breakdown is possible (they form a high-voltage "lock"). The cathode and its insulator are loose-fitted into the housing as a single assembly.

The gauze anode is inserted into the housing, and the anode ring is screwed in to fix the anode in place and ensure its electric contact with the housing. From the other end, the high-voltage cable is inserted into the cathode insulator and covered with the housing lid. The housing lid is then screwed to the housing.

The gauze anode is made of 12Kh18N10T stainless steel or an equivalent to ensure thermal and chemical resistance during plasma heating [99]. The gauze anode's copper ring is 2 mm thick.

The changes described in this section made it possible to rule out electric breakdown of the device's components and to reduce fluctuations of the gas-discharge current to 1.5%–2%, thereby ensuring that the device's parameters remain stable during long-term operation.

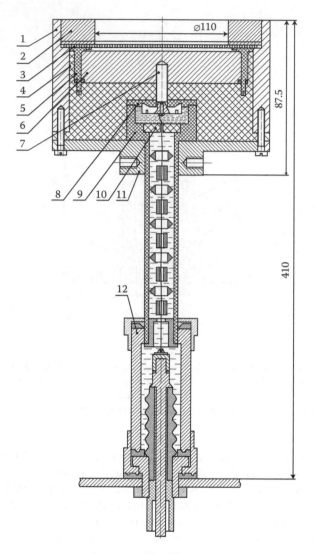

FIGURE 1.13 High-voltage low-temperature plasma generator: (1) housing, (2) anode ring, (3) gauze anode with copper ring, (4) cathode insulator, (5) PTFE gasket, (6) cathode, (7) screw, (8) rubber gasket, (9) high-voltage cable, (10) PTFE insert, (11) housing lid, and (12) high-voltage cable gland. Overall dimensions are in millimeters.

1.4 NEW DEVICES FOR GENERATING DIRECTED FLUXES OF LOW-TEMPERATURE OFF-ELECTRODE PLASMA

The studies and proposed design changes provide the basis for new devices to generate fluxes of low-temperature high-voltage gas-discharge plasma with set flux patterns. These devices may find application in contemporary diffractive optics to help solve a current problem: how to form complexly shaped optical microreliefs on convoluted surfaces.

1.4.1   MULTIBEAM GAS-DISCHARGE PLASMA GENERATOR

An ion source is known that is based on the injection of electrons from glow-discharge plasma through a small, fine-mesh opening to the plasma generator's cavity containing the main thin-wire anode [102]. Gas ionization induced by the electrons being injected into the cavity results in a dependent glow discharge with a hollow cathode, the plasma of which is the source of ions. Fast electrons oscillating inside the cathode cavity ensure that uniform plasma is generated at very low gas pressures.

But because the amount of ions extracted from the plasma is small—about 4%—the efficiency of this source is low.

A source of ribbon-shaped electron beams is known [103] that contains a cylindrical hollow cathode with a longitudinal slot in a sidewall, an anode with a metal mesh-covered emission window, and an accelerating electrode with a window for an electron beam to pass through, the hollow cathode of which is used to increase beam uniformity through repeated electron oscillation. But the cathode cavity's end walls induce a difference in the formation rate of ion–electron pairs near the walls and elsewhere in the cavity. This, in turn, results in current-density peaks along the beam edges and an increase in plasma concentration near the end walls.

A plasma electron source of ribbon beams is known [104] in which the inner end walls of a hollow cathode are covered with heat-resistant inorganic dielectric plates so that the particles in a plasma flux are distributed more uniformly.

But the device has several shortcomings: The emission window must be covered with a mesh. The device uses an accelerating electrode. Because of the anode design, a plasma flux can be formed only by one side of the hollow cathode's surface. The hollow cathode cannot simultaneously form several plasma fluxes of different cross sections.

Among the closest analogs to the generator proposed by this monograph is a device consisting of a hollow cathode, hollow anode with an emission mesh-covered emission window, high-voltage cable gland, accelerating electrode, and insulation, all arranged coaxially [105]. Between the accelerating electrode and the anode, there is a heat-resistant inorganic dielectric disk with a central hole whose diameter exceeds that of the anode's emission hole and is less than that of the accelerating electrode's hole. But because of its electrode arrangement, the device is not capable of forming several variously directed electron beams of different cross sections.

Described below are the design and operating principle of a multibeam gas-discharge plasma generator [106]. This device falls into the class of gas-discharge devices used for forming directed fluxes of low-temperature off-electrode plasma and generating convergent and divergent high-current ion–electron beams.

The device was designed to solve the problem of simultaneously generating several directed plasma fluxes of different cross-section shapes with a device having a higher discharge current and a lower electrode voltage, as well as a simpler design and operating conditions.

The problem was solved by equipping the multibeam gas-discharge plasma generator with a hollow cathode, hollow anode, insulation, and high-voltage cable gland. The hollow cathode attached to the base is coaxially installed into the hollow anode, and the cathode–anode insulation has a thickness of $\lambda < h < 3\lambda$, where λ is the mean free path of the electron in the gas-discharge plasma flux and h is the thickness of the dielectric material. We have patented this range of h values [106] because it was experimentally proved that outside the $\lambda < h < 3\lambda$ range, no discharge occurs in the proposed electrode system. The anode is covered with lids, thereby forming a sealed cavity. The coaxial holes in the surfaces of the anode, cathode, and insulation can be arbitrarily shaped in the form of a circle, ring, rectangle, triangle, and straight or curved slots. The insulation between the anode and cathode surfaces is made of PTFE or polystyrene (see Figure 1.14).

Figure 1.14 shows the design of the multibeam gas-discharge plasma generator. The device consists of a high-voltage electrode (1) that serves to supply power to a hollow cathode (2). The hollow cathode attached to the base (3) is coaxially installed into a hollow anode (4); cathode–anode insulation (5) is covered with a lid (6); the anode (4) is covered with lids (7, 8), thereby forming a sealed cavity; the surfaces of the anode, cathode, and insulation have coaxial S-shaped, circular, and rectangular holes.

Coaxial holes in the anode–insulation–cathode assembly are cut from the side of the lid (7) and the anode surface (4). The holes' shape must match that of the cross section of gas-discharge plasma. This results in an assembly incorporating a hollow anode and a hollow cathode. The number and shape of the holes are determined by the process conditions.

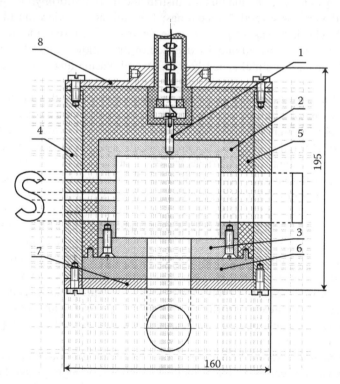

FIGURE 1.14 Multibeam gas-discharge plasma generator. Overall dimensions are in millimeters.

If a voltage of 0.3–6 kV is applied to the electrode system, the electric flux in the area of the holes is distorted. Without encountering any obstacles on their way, free electrons enter the cathode's cavity and ionize residual-gas atoms.

If an electron ionizes one or more atoms, a cloud of gas-discharge plasma appears in the cavity area. This cloud is an effective source of free electrons for generating a gas discharge in the anode-hole area. Because the space between the cathode and the anode is not sufficiently large for electrons to gain energy adequate to ionize the residual-gas atoms, ionization takes place outside the anode surface—that is, gas discharge takes place outside the device's electrodes. This makes it possible to obtain gas-discharge plasma fluxes measuring hundreds to thousands of milliamperes with voltages of 0.3–1 kV. Generating a gas-discharge plasma flux outside the device's electrodes eliminates the dependence of plasma parameters on the surface area (the loading effect). Depending on the process requirements, the cathode cavity can be of any form—that is, square, spherical, or rectangular. The main conditions for the occurrence of gas discharge are $H \geq 3\lambda$; $\lambda < h < 3\lambda$; and $Q\gamma > 1$, where λ is the electron's mean free path, H is the minimum length of the cathode cavity in the direction of ion movement, and h is the dielectric thickness in the cathode- and anode-hole areas. The critical requirement is $H \geq 3\lambda$ because if the inequality is not satisfied, no gas discharge occurs in the cathode cavity. A cathode–anode gap less than λ increases the possibility of a short circuit between them as a result of the mesh sagging during heating. Preventing the sagging warrants much more complex anode design. In contrast, small gaps between the cathode and anode can lead to arc discharge occurring between them, a condition that will permanently damage the gas-discharge device. A gap between the anode and cathode exceeding 3λ will lead to a parasitic discharge in it that falls into abnormal types of glow discharges, resulting in severe burning of the working surfaces of the cathode and anode.

The device generates several low-temperature plasma fluxes in the hole area simultaneously. The fluxes have similarly or differently shaped cross sections and the same or different flow directions.

1.4.2 GAS-DISCHARGE PLASMA FOCUSER

Among the closest analogs to the generator proposed by this monograph is a device consisting of a flat cathode, gauze anode, insulation between them, and a high-voltage cable gland [64]. The minimum space between the cathode and gauze anode is taken to be equal to the Aston dark space of a glow discharge. If the anode and cathode are made of a solid material, no gas discharge occurs between them. But because of its flat electrodes, this device cannot process convoluted surfaces because most plasma particles impinge on a convoluted surface at a certain angle, leading to nonuniform distribution of the energy absorbed by the surface of the material being treated and, therefore, nonuniform etching.

This poses the problem of generating a gas-discharge plasma flux whose particles bombard any point of a convoluted surface normally, with increased discharge current and decreased electrode voltage.

The problem can be solved by equipping a gas-discharge plasma focuser [107] with a cathode, gauze anode, insulation, and high-voltage cable gland. The cathode

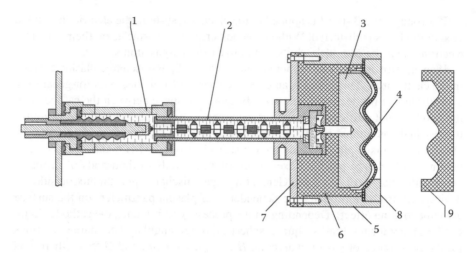

FIGURE 1.15   Gas-discharge plasma focuser.

and the gauze anode in this case have a curvature equal to that of the product being treated and are positioned at a distance of $15\lambda < h < 50\lambda$ from it (see Figure 1.15).

Figure 1.15 shows the design of the focuser. The high-voltage cable gland (1) and the high-voltage cable (2) supply power to the cathode (3). The cathode is installed coaxially into a hollow housing (5); insulation (6) is located between the cathode and anode; the housing is covered with a lid (7) and a gauze anode (4), thereby forming a gas-discharge electrode system. A ring (8) presses the anode to the housing, thereby providing an electric connection between the anode and the housing.

The device functions as follows. A product (9) whose surface curvature is in phase with that of the cathode and anode surfaces is installed into the cathode–insulation–anode assembly from the side of the anode's surface. If a voltage of 0.3–6 kV is applied to the electrode system, the electric flux in the area of gauze cells is distorted.

Without encountering any obstacles on their way, free electrons leave the anode and ionize residual-gas atoms. The length of an electron's mean free path falls within the range in which the electron gains in the accelerating field energy adequate for ionization. If the electron ionizes one or more atoms, a cloud of gas-discharge plasma appears in the anode–substrate area. This cloud is an effective source of chemically active particles contributing to the etching of the diffraction microrelief on a convoluted surface.

In the anode–cathode space, free electrons do not get to gain the energy sufficient to ionize the residual-gas atoms. Therefore, no discharge occurs outside the gauze cells. This makes it possible to obtain fluxes of gas-discharge plasma measuring hundreds to thousands of milliamperes with voltages of 0.3–1 kV. As in the case described earlier, generating a gas-discharge plasma flux outside the device's electrodes eliminates the dependence of plasma parameters on the surface area. The main conditions for the occurrence of gas discharge are $\lambda < h < 3\lambda$ and $Q\gamma > 1$. Other aspects of the device's operating principle are similar to those of the multibeam gas-discharge plasma generator.

When the focuser is operating, plasma fluxes form in the area of the anode's gauze cells, each moving strictly normally to the surface area of the related cell—that is, many variously directed microfluxes appear (see Figure 1.15). They collide with the surface at different angles, an effect corresponding to the different values of energy absorbed by the surface under treatment. Therefore, each surface point has its own etch rate, leading to the formation of a convoluted surface on the treated sample with a curvature equal to that of cathode and anode surfaces.

Creating a diffraction microrelief on a convoluted surface necessitates creating identical cathode and anode curvatures. In this case, with a mask applied to the convoluted surface, etching takes place only in the area not covered by the mask. Then, a diffraction microrelief is formed on the convoluted surface. But if the sample is positioned closer than 15λ from the anode, the gas discharge becomes unstable—that is, flashing takes place instead of stable burning, and surface etching stops. If the sample is positioned farther than 50λ from the anode, the plasma particles in this area do not move straight but in a manner similar to the particles of the positive column of a glow discharge. This chaotic movement significantly increases nonuniformity and decreases the etch rate, thereby causing the geometric parameters of microreliefs to deviate from their correct values.

1.5   CHAPTER SUMMARY

The theoretically and experimentally proven capability of high-voltage gas discharge to generate large-format fluxes of low-temperature plasma outside the gas-discharge device's electrodes enables us to propose several gas-discharge devices with unique capabilities. Our experiments proved that plasma fluxes generated by these devices contain charged particles that are distributed across the cross section with a high degree of uniformity. This makes for anisotropic, uniform etching of diffraction microreliefs over large surface areas. The proposed devices can be used for fabricating optical microreliefs on large-aperture diffractive elements.

1.5 CHAPTER SUMMARY

2 Methods for Quickly Measuring Surface Cleanliness

If diffractive optics is to continue developing, the existing range and functions of DOEs need to be expanded. This calls for more stringent cleanliness requirements for substrate surfaces since the quality of a diffraction microrelief depends on surface cleanliness. Stable and reproducible surface properties that meet the process requirements ensure that masks deposited on surfaces through the thin-film technique [108] demonstrate the desired parameters.

Molecular-level contamination on substrate surfaces cannot often be detected during fabrication of DOEs and micro- and nanoelectronic components but instead manifests itself only after the products have operated under extreme conditions over some time. This notably reduces their reliability. It is this problem that makes it critical that surface cleanliness be monitored as part of fabricating DOE microreliefs and micro- and nanostructures [16,17,109].

There are many methods and instruments for measuring the cleanliness of substrate surfaces [16–20,21,22,110–114], including instruments for analytical and quick measurements. Analytical measurement methods involve using equipment such as focused-beam and tunnel microscopes, Auger spectrometers, and spectrophotometers, and these yield reliable measurement data. But because of their costliness, high energy consumption, and low productivity [21,22,110], many of the methods are not suitable for quickly and accurately checking whether surface cleanliness meets the process requirements.

Quick measurements take much less time and use relatively simple, affordable instruments—an advantage that allows surface cleanliness to be measured directly at the workstation. But to give reliable information, such instruments need to be calibrated by the use of substrate surfaces with reference properties. This makes measurements much more complex because in this case individual reference surfaces are required for each type of contamination. The use of quick-measurement instruments also necessitates cleaning the indenter probe's surface with special techniques [18] or using special liquids as a source of information about the surface cleanliness of substrates [19,20]. These shortcomings highlight the need for improving the methods for quickly measuring surface cleanliness.

This section shows that for quickly measuring surface cleanliness, it is most convenient to use an instrument based on a tribometric method that involves measuring the coefficient of sliding friction. The instrument is convenient to operate, offers short measurement times, and, as will be discussed below, requires neither reference surfaces for calibration nor cleaning of the indenter probe's surface with special techniques.

2.1 OVERVIEW OF METHODS FOR QUICKLY MEASURING SURFACE CLEANLINESS

The key methods for quickly measuring the cleanliness of substrate surfaces are frustrated multiple internal reflection (FMIR) spectroscopy, the method of measuring the Volta potential, the tribometric method, and the wettability method. The latter method is further divided into the immersion method, the condensation method, and the wettability method involving the deposition of a liquid drop on a substrate surface.

2.1.1 METHOD OF FRUSTRATED MULTIPLE INTERNAL REFLECTION SPECTROSCOPY

When a light flux is multiply reflected from the substrate–contamination structure at the substrate surface, the reflected ray is attenuated because part of the flux passes from a denser medium to one that is less dense optically, and is absorbed by it. During the process, the contaminated surface of the test substrate contacts the FMIR element, or the test substrate itself serves as an FMIR element [113]. From this, it follows that the test substrate must ensure total internal reflection of the light ray in both cases, thereby ensuring the related optical parameters are achieved. But satisfying these conditions limits the method's sensitivity (a thickness less than one monolayer cannot be measured) and the types of substrate material and contamination that can be measured with the method.

The results of experimental studies in Figure 2.1 show that in the case of monomolecular films, surface cleanliness is measurable in a range limited to 10^{-7} g/cm^2 [113].

In addition, during the measurement procedure, the surface of the test substrate comes into contact with the FMIR element, and the surface of the FMIR element becomes contaminated. This necessitates cleaning the element for further

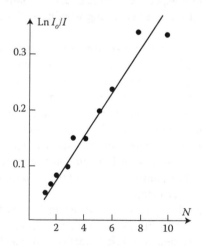

FIGURE 2.1   Intensity of the absorption band at $\lambda = 2{,}920$ cm^{-1} in the FMIR spectrum versus organic-film thickness [109]. N is the number of monolayers.

measurements, which is a serious drawback that impedes quickly measuring the cleanliness of substrate surfaces.

Thus, the optical properties of substrate materials and contaminants and a limited measurement range hinder the use of the FMIR method.

2.1.2 METHOD OF MEASURING THE VOLTA POTENTIAL

The method involves vibrating a point probe with a set amplitude above the test surface and determining the change in the probe's capacitance in response to contamination of the substrate surface. Numerical values of capacitance depend on the properties of both the contaminants and the dielectric material. According to Reference 115, the surfaces of solids show considerable concentrations of positive and negative charges that actively form the capacitive properties of the vibrating probe. This causes serious difficulties in interpreting the measurement results.

The probe's vibration causes a gas flow [114,116] whose particles collide with the test surface, thereby changing its properties. To prevent this effect, the Volta potential should be measured in an inert atmosphere or in a highly clean environment.

Because of its high sensitivity to the presence of surface charges, this method is highly sensitive to changes in the properties of initial substrate surfaces. As a result, the instrument's parameters need to be recalibrated for each new batch of substrates or gas composition in the working chamber. This suggests that the method is not optimally suitable for quickly measuring the cleanliness of substrate surfaces.

2.1.3 METHODS FOR EVALUATING CLEANING EFFICIENCY BASED ON WETTABILITY OF THE SUBSTRATE SURFACE

The simplest and most affordable methods giving accurate information about the amount of contamination on substrate surfaces are the immersion method and the condensation method [111,112], whose sensitivity is fairly high, that of the former being $1 \cdot 10^{-6}$–$1 \cdot 10^{-7}$ g/cm^2 and that of the latter being $1 \cdot 10^{-7}$–$1 \cdot 10^{-8}$ g/cm^2. But despite their simplicity, these methods suffer from serious drawbacks: they are suitable only for hydrophobic contaminants and contaminate the surface of the test substrate with residual liquid during the measurement.

The wettability method yields quantitative information about the level of contaminants on substrate surfaces such as information based on the values of the contact angle of a liquid drop [111,112] or the spreading rate of a liquid drop [16,17,19,20,117,118] on the test surface.

Measuring the contact angle Θ is technically difficult because the drop spreads over the surface in the process and changes its physicochemical properties as a result of dissolving hydrophilic surface contaminants in it. These drawbacks result in poor reproducibility of the readings obtained from instruments that are used in the method [117].

To prevent these drawbacks, References 16, 17, 19, 20, 117, and 118 propose measuring the cleanliness of substrate surfaces by determining the spreading rate of a liquid drop over the test surface as a function of the amount of contamination.

But for this method to be effective, the geometry and composition of the liquid drop must be closely monitored.

The volume of a liquid drop is easily measurable with calibrated cavities made of quartz or other chemically inert materials such as nickel–chromium alloys or kovar [119]. But preserving the properties of a liquid drop is difficult because water is a good solvent and its properties are actively influenced by the interaction of its structure and the surface of the reservoir containing the water. As a result, the reservoir's components have to be stored in special chambers and thoroughly cleaned each time the cleanliness of substrate surfaces is measured. According to Reference 18, if components precleaned to $1 \cdot 10^{-8}$–$5 \cdot 10^{-9}$ g/cm^2 are stored in a nitrogen hood, the surface becomes contaminated to a level of 1–$4 \cdot 10^{-7}$–$5 \cdot 10^{-8}$ g/cm^2. And measuring the cleanliness of substrate surfaces in the air atmosphere in a laboratory environment results in the surface of the reservoir's components being contaminated to 10^{-6} g/cm^2 (see curves 2 and 4 in Figure 2.2) [18].

As a result, when a quick-measurement instrument is operated in a laboratory environment, the inner surface of the distilled-water drop dispenser should be expected to be contaminated to 1–$6 \cdot 10^{-6}$ g/cm^2. The contamination corresponds to the beginning of calibration curves 1, 2, 3, and 4 in Figure 2.2.

Further, using a drop of distilled water as a probe limits the range of measurable substrates and contaminants, because when both the substrate and contamination are hydrophobic, the drop does not spread, and, conversely, when both are hydrophilic, the drop reaches its maximum spread on the surface and the amount of contamination has little effect on the drop. This causes some difficulty in obtaining accurate information from instruments used in the wettability method [19,20].

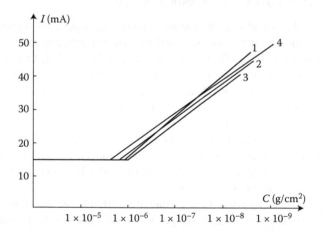

FIGURE 2.2 Calibration curve of the ICh-2 surface-contamination meter (tribometer): (1) curve from the meter's datasheet, (2) curve obtained experimentally for kovar samples cleaned by chemical polishing (contamination: IS-20 oil dissolved in trichloroethylene), (3) for kovar samples cleaned by redox annealing (contamination: IS-20 oil dissolved in trichloroethylene), and (4) for nickel samples cleaned by chemical polishing (contamination: oleic acid dissolved in trichloroethylene).

The major drawback of the wettability method's techniques is that the composition of the liquid drop contaminates the test surface (this necessitates additional cleaning). Despite the drawbacks described in this section, the wettability techniques are widely used for quickly measuring the cleanliness of substrate surfaces, owing to their low cost and the absence of a more efficient method.

2.1.4 TRIBOMETRIC METHOD

The tribometric method for measuring surface cleanliness is based on the dependence of the coefficient of static friction on the concentration of surface contaminants and makes it possible to measure concentrations up to $1 \cdot 10^{-9}$ g/cm^2 [18,109]. Figure 2.3 shows the schematic of a standard instrument used in this method.

This measurement is accomplished by determining the current that creates in the solenoid, at the moment the indenter probe starts moving, a traction force that is equal to the force of static friction. The galvanometer's relative readings correspond to the coefficient of friction being measured and therefore to the amount of contamination at the substrate surface.

Analysis of the instrument's design identifies its major faults. These are caused by poor reproducibility of the signal contacts' resistance because of the contacts' contamination and by friction forces from mechanically moving components, which are comparable with the indenter probe's friction force that is being measured over the test surface and which add to that friction force (see Figure 2.3). This causes significant error in the measurement of the coefficient of static friction, and eliminating the error is technically difficult.

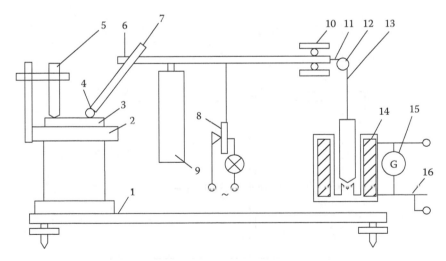

FIGURE 2.3 Schematic of the ICh-2 tribometer: (1) bed, (2) sample stage, (3) test substrate, (4) probe, (5) hold-down, (6) main axis, (7) indenter axis, (8) signal contacts, (9) load, (10) ball bearing, (11) link, (12) roller, (13) core, (14) solenoid, (15) galvanometer, and (16) potentiometer for measuring surface contamination.

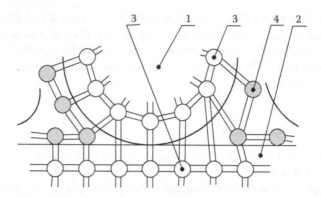

FIGURE 2.4 Schematic of indenter probe–substrate interaction at rest: (1) probe, (2) test substrate, (3) surface atoms, and (4) contamination atoms.

Besides the design faults and the measurement error caused by the voltage ripple at the solenoid [17], the measurement method itself suffers from the serious drawback that the indenter probe indents the contamination layer when the indenter probe is in stationary contact with the test substrate's surface [120]. This effect causes surface atoms of the indenter probe and those of the substrate to interact at the molecular level. Contamination atoms contribute to the process partially—that is, only along the perimeter of the tribometric contact spot (see Figure 2.4).

This interaction mechanism increases the measurement error. The error is well illustrated by the relationship between the friction force and the indenter probe's displacement in Figure 2.5 [121,122].

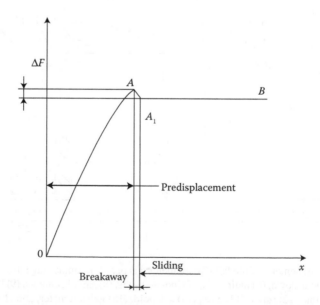

FIGURE 2.5 Friction force versus displacement of the indenter probe on the test surface.

It is the interaction between the surface atoms of the indenter probe and those of the substrate that explains the curve section AA_1. The value of ΔF characterizes the measurement error.

Another drawback of the instrument is the necessity of using a special cleaning technique for the indenter probe and the reference (calibration) surface [18]. A common cleaning technique consists of degreasing the indenter probe twice in boiling trichloroethylene or carbon tetrachloride for 10 min (a high-purity-grade or doubly distilled solvent must be used). To increase the accuracy and reproducibility of measurements, the indenter probe must be cleaned before each measurement or, alternatively, the contact point must be changed between the surface of the indenter probe and that of the test substrate.

The drawbacks discussed in this section make measurements more complex, reduce their reliability, and render the method of measuring the coefficient of static friction ineffective for quickly measuring the cleanliness of substrate surfaces. And because the accuracy of measuring surface cleanliness depends on the contamination level of the probe itself, chemical cleaning of the probe is undesirable: it results in the probe's surface being contaminated with the chemical itself.

From our analysis of the existing methods and devices for quickly measuring surface cleanliness, we can conclude that the tribometric method surpasses the other methods in terms of accuracy. But to overcome the method's drawbacks, the tribometer needs to be improved and an appropriate measurement procedure developed.

2.2   DESIGN CHANGES TO THE TRIBOMETER

This monograph proposes improving the ICh-2 tribometer and the tribometric method by

- Measuring the coefficient of sliding friction rather than the coefficient of static friction
- Replacing the indenter probe with a substrate-probe identical to the test substrate
- Substituting gravity force for the electromagnetic attraction force exerted by the inductive coil's armature on the indenter probe

These improvements appreciably simplified both the instrument's design and the procedure for measuring the cleanliness of substrate surfaces [70,71,123,124] and made it possible to prevent the effect of friction forces from the instrument's moving components on the measurement of the speed at which the substrate-probe slides along the surface. This, in turn, reduced the error of measuring contaminant concentrations. Besides, comparative analysis of the atomic bonds formed during the interaction between the surface of the indenter probe (the substrate-probe) and the test surface shows that using the coefficient of sliding friction as a criterion of surface cleanliness makes it possible to minimize the stress that the substrate-probe exerts on the test substrate in the area of their contact and to measure the electron bonds directly between the atoms contaminating these surfaces (see Figures 2.4 and 2.6) [71]. In the

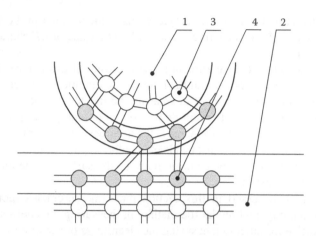

FIGURE 2.6 Schematic showing the interaction between the moving probe and the substrate: (1) probe, (2) test substrate, (3) surface atoms, and (4) contamination atoms.

latter case, it is possible to minimize the cohesive component of tribometric interaction, which results from the deformation of the interacting substrates' microreliefs, and to increase the adhesive component, which carries information about the contaminants held at the surface mainly by Van der Waals forces [120]. The measurement takes place in the segment A_1B (see Figure 2.5), and so the error due to ΔF is eliminated.

Figure 2.7 [70] shows the external view of the tribometer used in the method of measuring the coefficient of sliding friction.

A feature of the modified method is that it uses a criterion of surface cleanliness that is represented by the coefficient of sliding friction of two substrates cleaned

FIGURE 2.7 Tribometer for quickly measuring the cleanliness of substrate surfaces by measuring the coefficient of sliding friction.

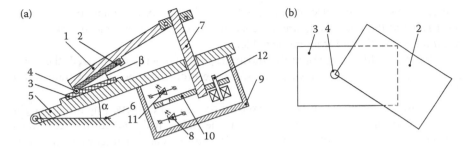

FIGURE 2.8 Arrangement of substrate holders (a) and substrates (b) in the tribometer: (1) substrate-probe holder, (2) substrate-probe, (3) test substrate, (4) point of tribometric contact between the two substrates, (5) test-substrate holder, (6) tribometer housing, (7) retaining rod of substrate-probe holder, (8) LED, (9) lightproof cover, (10) metal disk, (11) photodiode, and (12) rod-locking pin.

under identical conditions. This obviates the need for use of special techniques and materials for preparing the measuring probe's reference surface [18]. To make it possible to use one of the test substrates as a probe, this monograph proposes a sub-strate-holder assembly that provides a point contact between the substrates' polished surfaces. Figure 2.8a and b shows the arrangement of substrates in the substrate-holder assembly.

The measurement consists of placing a substrate whose concentration of con-tamination atoms and molecules is to be measured into the substrate holder inclined at an angle of α to the axis of abscissas. The angle's precise value is determined by the sum of weights of the substrate-probe holder and the substrate-probe, which add up to the gravity force determining the sliding behavior of the substrate. Then the substrate-probe is placed into its holder, whose surface is inclined at an angle of $\beta = 4\text{--}10°$ [123] to the test substrate's surface, providing a point contact between the two surfaces. Numerical values of the angles α and β are determined with a angle gauge with a scale division of 2′, designed for measuring inner and outer angles of parts. Systematic relative error for the two angles in the case under study measured $\Delta_\alpha = 0.11\%$ and $\Delta_\beta = 0.83\%$, respectively. The substrate-probe holder's structural dimensions depend on substrate dimensions.

In the initial position, the substrate-probe holder is moved to the topmost part of the test surface and is fixed in place with the rod-locking pin. Then the substrate-probe's working surface is brought into contact with the test substrate, providing a point of tribometric interaction between them. In the same position, a reference point is generated when the light ray from the light-emitting diode (LED) falls on the photodiode's surface through a hole in the metal disk. To generate pulses for measuring the substrate-probe's sliding speed τ, 10 1-mm holes spaced at 1 mm were made in the metal disk along a circle with a radius of 160 mm. To reduce systematic instrument error in directly measuring the value of τ, the holes' diam-eters and spacing were measured with an MK micrometer of 102 model (compliant with the GOST 6507-90 standard) with a scale division of 0.01 mm. The relative error amounted to $\Delta_d = 1\%$ and $\Delta_l = 1\%$, respectively. The components that serve to generate measurement pulses are enclosed with a lightproof cover and fixed in place: the LED and the photodiode are attached to the cover and the metal disk, to the retaining rod of the substrate-probe holder.

When the rod-locking pin of the substrate-probe holder is disengaged, the point of tribometric interaction between the substrate-probe and the test substrate slides down.

Because the metal disk is rigidly fixed on the axis of the retaining rod, the light flux flashing onto the photodiode through the holes in the metal disk generates current pulses in the photodiode circuit. The pulse time τ characterizes the sliding speed and is proportional to the numerical value of the coefficient of sliding friction μ.

The pulse group generated in the photodiode circuit is then processed by the electronic circuit shown in Figure 2.9.

The circuit is based on the *DD*1 (AT89S52) microcontroller, which is used for the following purposes:

- Measuring the pulse time of a light flux with the pulses from a clock generator
- Calculating the mean of three pulses of a light flux, as well as the deviation from the entered reference values
- Displaying the measurement results on the seven-segment LED display
- Interfacing with the external computer

The microcontroller is run by the program stored in its internal memory. The operation algorithm is shown in Figure 2.10.

The circuit functions as follows. The light flux generated by the LED VD7 passes through the holes in the rotating metal disk and generates light pulses that are incident on the photodiode VD5. Concurrently, a swing from 1 to 0 occurs on the collector of the transistor VT2, while the absence of light-flux pulses results in a swing from 0 to 1. The signal from the collector of the transistor VT2 arrives at the microcontroller's interrupt input. When the light-flux pulse starts acting on the photodiode, the microcontroller starts measuring the pulse time of the light flux by completing the time interval with precision pulses of the quartz generator with a pulse-repetition period of $T_{pls} = 0.083(3)$ μs. (The pulse counter's systematic relative error in the case under study was $\Delta_\tau = 0.016\%$.)

A display-segment control signal is generated on the terminals of the port P1. Then this signal arrives at the input of the integrated circuit DD4 (K1533AP5), designed to match the microcontroller to the display. This signal is fed from the outputs of the integrated circuit DD4 to the display HG1.

At bits 0–4 of the port P0, a control signal is generated for display bits and for keyboard scanning. This signal arrives at the inputs of the integrated circuit DD3 (K1533AP3), which serves as a signal shaper that inverts the signal and then transmits it to the display and the keyboard. The port P2 serves as an input for the keyboard-scanning signal. The diodes VD1–VD4 (KD521) are designed to pass the keyboard-scanning signal from the port P0 to the port P2 through closed keys and to prevent it passing in the opposite direction. The display HG1 uses an ALS318G

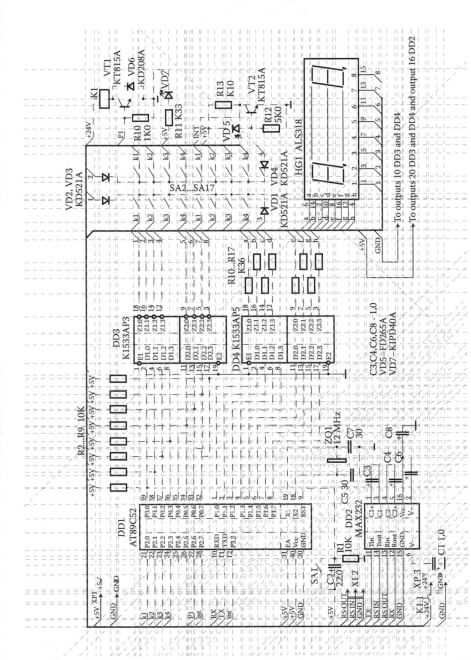

FIGURE 2.9 Circuit diagram of the tribometer.

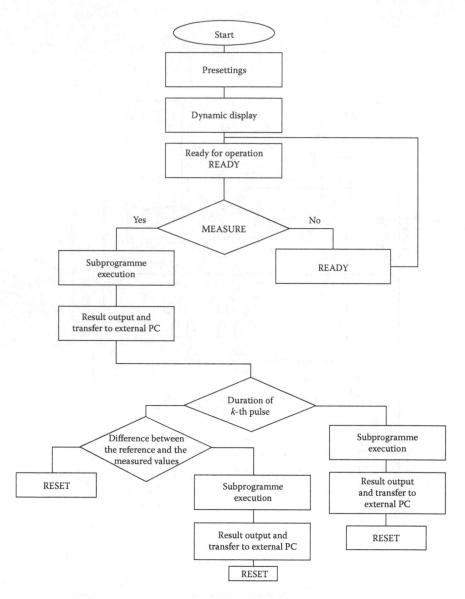

FIGURE 2.10 Software algorithm of the DD1 (AT89S52) microcontroller.

10-digit LED display. The keyboard has 16 keys, with the function of each determined by the software.

The microcontroller DD1 uses a universal asynchronous receiver/transmitter that supports the RS-232C protocol, thereby providing a connection to a PC. For this connection, the integrated circuit DD2 is used, which exchanges data with the external computer.

The transistor VT1 operates in the switching mode, which is needed to control the solenoid coil K1 of the rod-locking pin for the substrate-probe holder in the upper position (in the topmost part of the test substrate). Following a 1-s delay after the Measure key is pressed, the transistor VT1 gets closed by a logic-zero level signal from output 27 of the microcontroller DD1, and the substrate-probe holder moves to the lower position down the test substrate's surface.

The components SA1, C2, and R1 form an automatic/manual reset circuit. An automatic reset occurs each time the device is switched on. When the device is operating, a manual reset can be accomplished by pressing the SA1 button.

The resistors R2–R9 are used for adjusting the terminals of the port P0 to the logic-1 level.

The diode VD6 is a damper diode required to protect the collector–emitter junction of the transistor VT1 from voltage surges that occur in the solenoid coil K1 at the time of the transistor's switching.

As Figure 2.10 shows, the software algorithm starts with entering input data in the microcontroller's internal registers. The software dynamically displays measurement and calculation data and key words (such as *Ready* when the system is ready for operation).

Pressing the Measure key invokes a subprogram for scanning the sensor's status. If a swing from 1 to 0 is detected, the pulse time of the light flux is measured in the units of the quartz generator's timing pulses until a swing from 0 to 1 occurs. The measured value is converted from binary code to seven-segment code and is transmitted to the display circuit. If necessary, the measured value in binary code is also transmitted to the external computer's memory.

For calculating the duration of the k-th pulse and the difference between the reference and the measured values, commands are entered from the keyboard, causing the appropriate subprograms to be executed. Calculated values are transmitted to the display circuit and the external computer's memory.

Pressing the Reset key resets the microcontroller's internal registers and clears its RAM; then the operation starts afresh with the initial settings in accordance with the algorithm.

The total systematic error is determined by the algebraic sum of Δ_d, Δ_l, Δ_α, Δ_β, and Δ_τ [125] and amounts to $\Delta_c = 2.96\%$.

To increase the instrument's accuracy and resolving power, an optical grating [126] was placed in the hole area of the metal disk (see Figure 2.8) to split the light flux generated by the LED in the hole area into a set of secondary light sources. The sources should be separate in the area of the photoreceiver's aperture and should yield a picture showing the maxima and minima of light-flux intensity. At a distance of R from the grating, light fluxes from the secondary sources merge, making it next to impossible to register the fluxes emerging from individual slits

To register fluxes from individual slits, numerical values of the grating constant and the distance R to the photoreceiver must be determined.

In diffraction at one slit, 90% [1,2] of light-flux intensity spreads toward the observation point within the angle $\alpha = 2\lambda/b$, where λ is the wavelength of light and b is the slit width. The maxima of the second, third, and k-th orders do not have a notable effect when added to the first-order maxima; therefore the minimum allowable grating constant is determined from the condition that first orders of diffraction from adjacent slits not intersect:

$$T = \frac{2\lambda R}{\sqrt{b^2 - \lambda^2}}.$$

Therefore, if the light source is a semiconductor LED with a wavelength of $\lambda = 0.63\ \mu m$ and the grating's slit width b equals $20\ \mu m$, the expression for the grating constant changes to $T = 0.063R$; that is, the grating constant has a directly proportional dependence on the distance between the grating and the radiation receiver's screen. For the instrument to operate properly at $R = 1$ mm, the grating constant should be $63\ \mu m$.

Our comparative analysis of the metal disk's hole size and the grating constant shows that using an optical grating increases the instrument's resolving power by a factor of 16 by splitting the substrate-probe's sliding path into $63\ \mu m$ long sections, within which the surface cleanliness of the test substrate is measured. The duration of the pulses obtained depends on the coefficient of sliding friction and is recorded by the control and processing unit as a criterion of cleanliness for the test surface.

The external computer's display shows the pulse times obtained from the measurement. The keyboard is used to operate the unit and select light-flux times for any hole in the metal disk or the arithmetic mean of these values. The software converts pulse times proportional to the substrate-probe's speed to numerical values representing the concentration of contaminants at the test substrate's surface.

The instrument is powered from the 220 VAC, 50 Hz mains, weighs 3 kg, and measures $320 \times 200 \times 190$ mm. Measurements with the instrument must be taken in a room that meets vacuum-cleanliness requirements.

To summarize, this section discussed changes to the tribometer, consisting of using the coefficient of sliding friction for measuring surface cleanliness; using as a probe any substrate from a batch cleaned under identical conditions; designing suitable substrate holders allowing the probe to move by gravity and providing a tribometric point contact between the substrates; designing a circuit for measuring the time pulses proportional to the coefficient of sliding friction; and using an optical grating as a component for generating time pulses and improving the instrument's resolving power.

2.3   OPERATING REGIMES AND PARAMETERS OF THE TRIBOMETER

To determine appropriate operating regimes for the tribometer, we carried out an experiment to obtain curves that demonstrate the relationship between the pulse time τ and the process parameters: the angles α and β, the number of passes made by the substrate-probe and of cleanliness measurements, and the time that the substrates are exposed to the air. To verify the reliability and reproducibility of the results, each regime was reproduced at least 10 times, with the parameter spread no greater than 10%. For the results of our statistical analysis of the experimental data (calculation of random error), see Appendix A. Points in the graphs represent the mean of the pulse times for measurements taken on 10 samples.

According to References 70 and 71, in the case of assessing a substrate surface with minimum overall dimensions, measurement pulses are generated by 3 holes, while in the case of maximum overall dimensions, the number increases to 10. The hole diameter (or the slit width of a grating) and the substrate-probe's sliding speed relative to the test surface determine the time during which the light from the LED is incident on the active surface of the opto-isolator's photodiode—that is, the pulse time τ displayed on the tribometer's display. Because τ provides information about the coefficient of sliding friction and, therefore, about surface cleanliness, a shorter pulse time corresponds to a more contaminated surface and vice versa.

The selected angle β at which the substrate-probe is inclined to the test surface determines the shape of the tribocontact spot during the measurement of surface cleanliness. The surface shape of this tribocontact can have a significant effect on the measurement. To determine the nature of this effect, we have investigated the relationship between the pulse time and the angle β (see Figure 2.11).

The curves were obtained from two substrate batches with different levels of surface cleanliness. Analysis of curves 1 and 2 shows that they have a low extremum at $\beta = 5.5°$. As the angle β increases, the surface area of the contact between the substrate-probe and the test surface decreases and the coefficient of sliding friction is, to a larger extent, determined by the properties of the layer of organic contaminants adsorbed at the substrate surface. The manner in which the curves change shows that in the range $4° \leq \beta \leq 6°$, the angle β has an insignificant effect on the duration of the light pulse and that the effect is negligible for purposes of configuring the tribometer.

But at $\beta = 5.5°$, the extremum point indicates the optimal surface area of tribocontact between the substrate-probe and the test surface, and the coefficient of sliding friction is, to a larger extent, determined by the properties of the layer of organic contaminants adsorbed at the substrate surface.

In the range $\beta > 5.5°$, the contact area decreases to the extent that the substrate-probe indents the layer of organic contamination and the mechanical interaction of the substrates begins to prevail in the sliding process, resulting in a measurement

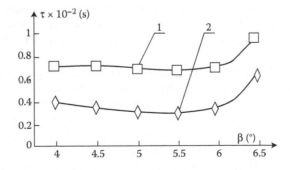

FIGURE 2.11   Pulse time τ versus the angle β at $\alpha = 45°$: (1) substrate series with the initial contamination level corresponding to $\tau = 0.7 \cdot 10^{-2}$ s and (2) to $\tau = 0.4 \cdot 10^{2}$ s.

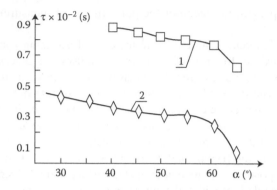

FIGURE 2.12 Tribometer readings versus the angle of the test-substrate holder at $\beta = 5.5°$: (1) substrate series with the initial level of contamination corresponding to $\tau = 0.9 \cdot 10^{-2}$ s and (2) to $\tau = 0.45 \cdot 10^{-2}$ s.

error identical to that of the method described in Reference 109. In this case, the substrate-probe can be likened to a cutting tool with a cutter whose contact area decreases as the angle β increases.

Therefore, to obtain reliable results about the level of surface contamination in practice, the angle β should not exceed 6°.

The sliding process depends on the angle of inclination α of the test-substrate holder. The relationship between the tribometer's readings and that angle is shown in Figure 2.12 [127].

The manner in which curves 1 and 2 change shows that the sliding process consists of two mechanisms: (1) the sliding of the substrate-probe on the test surface and (2) the free fall of the substrate-probe. The first mechanism occurs at $\alpha \leq 60°$ because the light-pulse time in this section changes by no more than 33% for both curves. This is due to the effect that the interaction between the substrate-probe's surface and the molecules of organic contaminants (in this range of α values) has on the coefficient of sliding friction. At $\alpha > 60°$, the adhesion strength decreases between the substrate-probe and the test surface, and $\alpha = 90°$ virtually places the indenter under free-fall conditions. It is this that explains an increasing acceleration of the substrate-probe in the range $60° < \alpha < 65°$. With these considerations in mind, we selected the range $30° \leq \alpha \leq 60°$ to take precision measurements. As this range is virtually free from the relationship $\tau = f(\alpha)$, the tribometer's readings within the range are stable.

Selecting $\alpha < 30°$ is inadvisable since the projection of the gravity force acting on the probe becomes comparable with the friction force or even falls below it, thereby stopping the measurement of substrates whose surfaces are close to technologically clean (for a definition of the term *technologically clean surface*, see Chapter 3).

At $30° \leq \alpha \leq 60°$, random error in measuring τ did not exceed 3% and at $4° \leq \beta \leq 6°$, 2% (see Tables A.5 through A.8 in Appendix A). The total random error determined by using the rules of geometric addition [125] amounted to $\Delta_{rndm} = 3.6\%$. The total instrument error (the error introduced by the instrument itself) amounted to $\Delta_{ttl} = 4.66\%$.

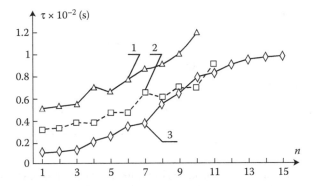

FIGURE 2.13 Tribometer readings versus the number of passes along one path for three substrate batches with different levels of contamination: (1) substrate series with the initial contamination level corresponding to $\tau = 0.5 \cdot 10^{-2}$ s, (2) to $\tau = 0.3 \cdot 10^{-2}$ s, and (3) to $\tau = 0.1 \cdot 10^{-2}$ s.

Assessing cleaning efficiency with tribometric methods requires knowing the extent of the effect that the indenter has on the properties of the test surface. For this purpose, Figure 2.13 shows the results of investigating the relationship between the pulse time and the number of passes made by the substrate-probe along a path on the test surface, for three substrate batches with different levels of contamination.

The manner in which curves 1, 2, and 3 change shows that, in the range $n \leq 3$ in the case of samples from all three batches, τ is virtually independent of the number of passes made by the substrate-probe—that is, that the coefficient of sliding friction in this case is determined entirely by the concentration of organic-contaminant molecules at the test surface. But at $n > 3$, the substrate-probe begins to slow down (the pulse time increases), which is an effect attributable to the intensified mechanical interaction between the two surfaces—as n increases, the number of the contamination molecules scraped off by the indenter within its sliding path increases as well. In the case of $n > 15$, interaction between the substrate-probe and the test surface takes place, which is confirmed by the saturation of curve 3 in this section. It follows that, to obtain accurate information on surface cleanliness, a substrate-probe must not be used more than three times along the same path.

Monitoring surface quality as part of fabricating a diffraction microrelief requires information about the cleanliness of the entire substrate surface. To satisfy this requirement, we have investigated the relationship between the value of τ and the number of measurements of the test substrate's entire surface area with one indenter (see Figure 2.14).

As the graph in Figure 2.14 shows, in the case of substrates with the initial contamination level corresponding to $\tau = 0.4 \cdot 10^{-2}$ s (curve 1), the substrate-probe can be used up to eight times because the tribometer's readings in this case change by no more than 12%, thereby allowing multiple use of the substrate-probe. As the contamination level decreases (τ increases), the mechanical interaction of the substrate surfaces begins to prevail, leading to the strict dependence of the tribometer's readings on the number of measurements (curves 2 and 3), and this agrees with the results shown in Figure 2.13. Another important feature of these curves is that they show

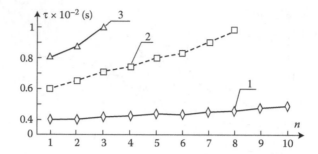

FIGURE 2.14 Value of τ versus the number of times the cleanliness of test substrates was measured with one indenter: (1) substrate series with the initial contamination level corresponding to $\tau = 0.4 \cdot 10^{-2}$ s, (2) to $\tau = 0.6 \cdot 10^{-2}$ s, and (3) to $\tau = 0.8 \cdot 10^{-2}$ s.

that even with a single measurement the tribometric method not only yields accurate results about the level of surface cleanliness but also allows substrates to be classified by this criterion. With computer-aided analysis, a substrate-probe can be repeatedly used for measuring the cleanliness of the entire surface area as well as surfaces with different levels of contamination.

Figure 2.15 shows the results of investigating the relationship between the instrument's readings and the time that the substrates are exposed to the air.

As the graph shows, exposure of cleaned substrates to the ambient environment for more than 5 min leads to an increase in their contamination level by 11%, 22%, and 28%, respectively; and the higher the level of contamination after a substrate has been cleaned, the more quickly its surface cleanliness changes, and the less the value by which it changes. The rate at which a clean surface adsorbs contaminants (curve 3) is 5.3% per minute; therefore, to obtain as accurate measurement results as possible, the time that the surface is exposed to the ambient environment should be limited.

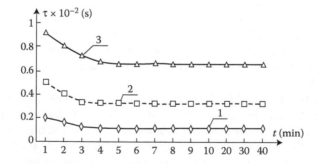

FIGURE 2.15 Tribometer readings versus the air-exposure time of substrates: (1) substrate series with the initial contamination level corresponding to $\tau = 0.2 \cdot 10^{-2}$ s, (2) to $\tau = 0.5 \cdot 10^{-2}$ s, and (3) to $\tau = 0.9 \cdot 10^{-2}$ s.

The minimum time in the experimental graph amounts to 60 s (see curves 1, 2, and 3 in Figure 2.15). The tribometer's simple design and ease of use allow one measurement to be taken within 15–20 s [70] and thus help obtain reliable results.

Because a monomolecular contamination layer is adsorbed from the air atmosphere within a period of 10^{-4} s [21,108,128], the measurement will always include an error caused by this effect (the error is inherent in any method for quickly measuring surface cleanliness under air-exposure conditions). Determining the actual value of surface cleanliness directly after cleaning the surface with a low-temperature plasma flux in the vacuum system can be accomplished by extending the related adsorption sections of curves 1, 2, and 3 in Figure 2.15 until they intersect the axis of ordinates. Using software to process the readings and increasing the number of measurement points through the use of an optical grating allow the actual values of surface cleanliness to be obtained in a single pass of the substrate-probe.

To summarize, this section discussed how the parameters and operating regimes of the tribometer used in the method of tribometric interaction between two substrates can be optimized for quickly measuring surface cleanliness. We have determined the operating regimes and parameters at which the surface cleanliness of substrates can be measured as accurately as possible.

2.4 DETERMINING THE EVALUATION CRITERION OF A TECHNOLOGICALLY CLEAN SURFACE

In carrying out our experimental investigations, we have found that the tribometer's readings depend not only on the coefficient of sliding friction but also on the geometric arrangement of the test substrates (see Figures 2.11 and 2.12). As a substrate surface is highly sensitive to any change in the angle α of the substrate holder (see Figure 2.12), an optimal range must be found for the angle α, within which the substrate-probe will not deform the surface structure of the test substrate. Thus, to obtain reproducible measurement results, we should analytically link the numerical values of the coefficient of sliding friction μ to the angles α and β and determine how they depend on the measured value of τ and on the concentration of contaminants at the substrate surface. The following considerations will help us find a solution to this problem.

Assuming that the substrate-probe moves along the test substrate with acceleration a, three forces act in the area of interaction between the substrates: the gravity force mg, which causes the substrate to move; the support reaction force N; and the sliding friction force F_{fr} (see Figure 2.16). The relationship $\mu = f(\alpha,\beta)$ can be expressed as [126]

$$\mu = \frac{g\cos[90-(\alpha+\beta)]-\alpha\cos\beta-g\cos\alpha\sin\beta}{g\cos\alpha\cos\beta}$$
$$= \frac{g\cos[90+(\alpha+\beta)]}{\cos\alpha\cos\beta}-\frac{a}{\cos\alpha}-tg\beta = tg\alpha-\frac{a}{g\cos\alpha}. \tag{2.1}$$

Increasing the angle α yields its critical value at which the movement of the substrate-probe will be determined not only by the properties of the test surface but also

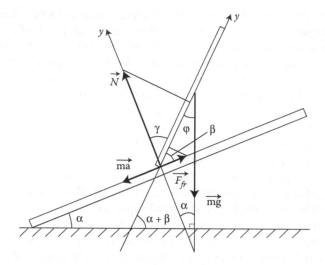

FIGURE 2.16 Distribution of forces that result from the contact of the substrates: (α) the angle of the test substrate to the horizon and (β) the angle between the surfaces of the contacting substrates.

the projection of gravitational acceleration on the direction in which the substrate-probe is moving. Such an effect will seriously corrupt the measurement results. This necessitates satisfying the inequality $(\alpha + \beta) < 90$.

In contrast, in the case of tribocontact between technologically clean surfaces, at a certain value of the angle α, the substrate-probe stops sliding over the test surface because the equation $\mu = \mu_{sf}$ is satisfied, where μ_{sf} is the coefficient of static friction [124].

Satisfying this equation allows the numerical value of μ_{sf} to be used as a criterion for evaluating the cleanliness of substrate surfaces as technologically clean as it allows the instrument's parameters to be calibrated against the reference values of μ_{sf}.

As an example, for silicon dioxide with an atomically clean surface, the coefficient of static friction in vacuum at a pressure of 10^{-7}–10^{-8} Pa equals 0.5, while in the air it is 1. This is attributable to the chemisorption of water molecules at the surface [129].

Thus, a technologically clean surface exposed to the air is invariably characterized by the presence of adsorbed water molecules and the absence of other impurities, and expression (2.1) at $\mu = \mu_{sf}$ changes to $tg\ \alpha = 1$. It follows that the technologically clean surface of a silicon dioxide substrate corresponds to the angle $\alpha = 45°$. For other materials, the criterion for evaluating a surface as technologically clean is determined similarly by substituting the appropriate value of μ_{sf} into Equation 2.1 and determining α. The above value of the angle α agrees with the experimental curves in Figure 2.12, since it falls within the ranges where the dependency $\alpha = f(\tau)$ is insignificant. From this it follows that $\alpha = 45°$ is suitable for a broad range of both substrate materials and types of contamination.

The manner in which curve 2 (Figure 2.12) changes shows that the sliding process depends on two mechanisms: the sliding of the substrate-probe over the test surface and the free fall of the substrate-probe. The first mechanism takes place at $\alpha \le 60°$. The changes in the light-pulse time in this section of the curve are linear and do not exceed 33%. As the changes are caused by the increase in the substrate-probe's potential energy, they are easily computable and used as an adjustment coefficient in calculating the coefficient of sliding friction. In the section $\alpha > 60°$, the adhesion strength decreases between the substrate-probe and the test surface, and $\alpha = 90°$ virtually places the substrate-probe under free-fall conditions. It is this that explains an increasing acceleration of the substrate-probe in the range $60° < \alpha < 70°$. These considerations confirm our selection of the range $30° \le \alpha \le 60°$ with the weak dependence $\tau = f(\alpha)$ as a range enabling the tribometer to yield stable and reliable readings.

A major difficulty in indirect measurement methods is how to convert instrument readings to numerical values that represent the concentration of atoms and molecules contaminating the substrate surface. When the coefficient of sliding friction is used as a criterion of surface cleanliness, this conversion can be accomplished taking into account the following assumptions: In our experiment, the sliding speed was expressed in relative units from zero to one and was an integral indicator. At a sliding speed of $V_{sl} = 0$ relative units, the contaminant concentration C_d was taken to be equal to the minimum detectable concentration corresponding to a technologically clean surface, 10^{-9} g/cm^2 [18], at which the measurement of μ_{sf} is taken [129]—that is, $C_d \le C_{d\,\min}$. At $V_{sl} = 1$ relative unit, $C_d = C_{d\,\max}$ (where $C_{d\,\max} = \rho h$, ρ being the density of organic contaminants and h the thickness of the contamination layer) is the impurity concentration corresponding to the presence of a contamination monolayer at the surface, $C_{d\,\max} = 10^{-7}$ g/cm^2 [109,121].

Because substrates undergo rough cleaning by immersion in a washing liquid, they have uniformly contaminated surfaces. Final cleaning uses a plasma flux with particles uniformly distributed across its cross section (see Chapter 1). Given the uniformity of the initial contamination, we assume that the final cleaning of substrates does not alter the manner in which the atoms and molecules contaminating their surfaces are distributed.

To determine the actual values of $C_{d\,\min}$ and $C_{d\,\max}$ corresponding to the threshold values of $V_{sl} = 0$ and $V_{sl} = 1$, measurements were taken on two substrates, one of which carried a monolayer of organic contamination (vacuum oil) and the other was cleaned to the maximum extent possible. The thickness of the monomolecular contamination layer was assessed with a P4-SPM-MDT scanning probe microscope, and $C_{d\,\min}$ was taken to be equal to 10^{-9} g/cm^2. As the surface was contaminated uniformly, the substrate-probe's sliding speed was proportional to the concentration of contaminants. Given these considerations, the following expression was used to convert the relative values of the sliding speed to the values of the surface concentration of organic contaminants, expressed in grams per square centimeter:

$$C_d = (1 - V_{sl})C_{d\,\min} + C_{d\,\max}V_{sl} = \left(1 - \frac{\tau_0}{\tau}\right)C_{d\,\min} + C_{d\,\max}\frac{\tau_0}{\tau}, \qquad (2.2)$$

where C_d is the concentration of organic contamination (g/cm²), and τ and τ_0 are, respectively, the pulse time measured at the tribometer's output and the pulse time for the substrate with a monomolecular contamination layer.

Deriving C_d from expression (2.2) is an indirect measurement. The error of method calculated as described in Reference 125 was no greater than 14.8%.

Thus, the proposed method for tribometric measurement of surface cleanliness makes it possible to calibrate the instrument without using surfaces with reference properties.

2.5   TRIBOMETRIC EFFECT OF THE SUBSTRATE-PROBE ON THE STRUCTURE OF THE TEST SURFACE

Tribometric interaction may cause mechanical damage to the structure of the test surface. The damage is undesirable because nanotechnology requires minimizing surface defects (structural damage) as the field develops. This poses the challenge of investigating the tribometric effect of the substrate-probe on the structure of the test surface to determine whether interaction between the two surfaces results in mechanical damage to the test surface.

For this purpose, the test surface was examined with a Solver PRO-M scanning probe microscope [126,130]. The results of the examination are shown in Figures 2.17 and 2.18. Comparative analysis of the images in Figures 2.17b and 2.18b shows that a normal stress exceeding 0.3 kg during tribometric interaction leads to mechanical damage to the surface structure of the substrates. For the substrate-probe's path, the profilogram shows traces of this damage—narrow scratches, sharp peaks sheared off, and many shorter and narrower peaks formed on the initial surface. On the whole, this results in a smoother surface relief.

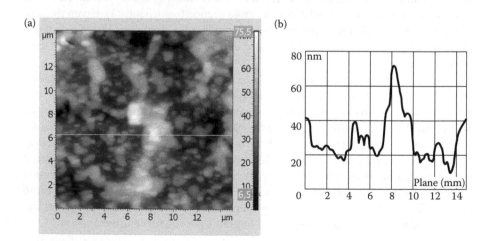

FIGURE 2.17   View of the test substrate's initial surface obtained with NT-MDT's Solver PRO-M scanning probe microscope: (a) view obtained in forced scanning mode (the white line indicates the trace of the microscope's cantilever) and (b) profilogram.

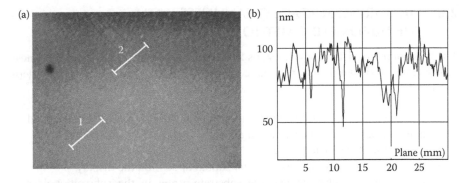

FIGURE 2.18 Structure of the test surface, obtained with Solver PRO-M: (a) photomicrograph of the area of the substrate-probe's pass for a stress of 0.3 kg (1) and 0.5 kg (2), magnified (×300), and (b) profilogram from region 2.

With the normal stress exerted by the substrate-probe reduced to 0.3 kg or less, the properties of the test surface virtually do not change—that is, its structure remains the same as the initial substrate (see Figures 2.17a,b, and 2.18a, region 1).

An atomic-force microscope (AFM) image showing the surface of the test substrate after tribometric contact at the angle $\beta > 6°$ revealed that in this case the sharp side edges of the substrate-probe's polished ends interact with the test surface. Figure 2.19 shows scratches resulting from the contact.

To summarize, when measurements are taken in the range $\beta = 4°–6°$ at $N \leq 0.3$ kg, the tribometric effect of the substrate-probe on the test surface does not cause mechanical damage to the test substrate's surface structure.

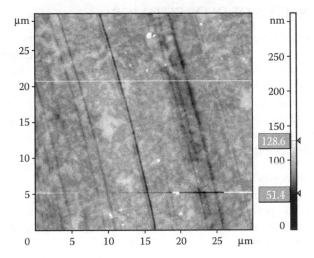

FIGURE 2.19 View of the test surface inclined at an angle exceeding 6° (the area of the substrate-probe's path).

2.6　MEASURING SURFACE CLEANLINESS WITH THE TRIBOMETRIC METHOD

The procedure consists of several steps for measuring the technological cleanliness of substrates or the amount of residual contamination after final cleaning. The steps are as follows:

- Position the test-substrate holder at an angle of $\alpha = 45°$ to the horizontal axis and position the substrates at an angle of $\beta = 5.5°$ to each other.
- Toggle the Reset switch to On.
- Press the Reset key. The word *Hello* should appear on the display.
- Place the test substrate and the substrate-probe in the substrate-holder assembly.
- Put the substrate-probe holder in the initial position.
- Press the Measure key.
- Record the tribometer readings.
- Press the Reset key.
- Remove the test substrate and place it into the desiccator.
- Repeat steps 4–9 to measure the surface cleanliness of other test substrates. (The total measurement time does not exceed 1 min.)

For a technologically clean surface, the reading of 0 relative units should correspond to 10^{-9} g/cm².

This procedure makes it possible to quickly obtain experimental data on the efficiency of the techniques used for rough and final cleaning as part of fabricating microstructures and microreliefs on DOEs.

2.7　A CLEANLINESS ANALYZER BASED ON ANALYSIS OF DROP BEHAVIOR

The methods of tribometric measurement discussed above are ideal for a production environment thanks to their ease of use and the low cost of equipment used [131]. For a scientific laboratory environment, a promising line of work is research on the possibility of using devices that are based on analyzing the dynamic state of a liquid drop deposited on the test surface [20,132]. An advantage it offers is the capability of analyzing a large surface area more quickly in comparison with the scanning probe microscope and the focused-beam microscope.

Liquid drops can form when:

- The liquid slowly flows from a small hole
- The liquid flows off the surface edge
- The liquid is sprayed or emulsified
- Vapor condenses on hard nonwettable surfaces or on condensation centers in a gas environment

The liquid surface tends to contract to the minimum surface area. In any case, three phase interfaces should be considered: gas–liquid, liquid–solid, and gas–solid.

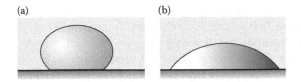

FIGURE 2.20 Different drop shapes on the surface of a solid in the case of nonwetting (a) and wetting (b) liquids.

The behavior of the liquid drop will depend on the surface tension (specific values of the free surface energy) in these phase interfaces. The surface tension in the gas–liquid interface will tend to give the drop a spherical shape. This will happen if the surface tension in the liquid–solid interface exceeds that in the gas–solid interface (see Figure 2.20a). In this case, the drop's contracting to a spherical shape leads to a reduction in the surface area of the liquid–solid interface and a simultaneous increase in the surface area of the gas–liquid interface. Under these conditions, the surface of the solid is observed to be unwettable with the liquid. The drop's shape will be determined by the resultant force of the surface tension and the gravity force. If the drop is large, it will spread over the surface, and if it is small, it will tend to contract to a ball shape [133].

If the surface tension in the liquid–solid interface is less than that in the gas–solid interface, the drop will take a shape that allows the surface area of the gas–solid interface to be reduced—that is, the drop will spread over the surface (see Figure 2.20b). In this case, the wetting of the solid is observed.

To describe the wetting of the solid in quantitative terms, the balance of forces should be considered that act on the line element formed by the intersection of three phase interfaces: (1) gas, (2) liquid, and (3) solid (see Figure 2.21).

In the first case, the force of attraction between water molecules exceeds the force of their interaction with the molecules of the solid, resulting in the liquid's taking a shape close to that of a ball (see Figure 2.21a).

In the second case, the reverse is true: the force of interaction between water molecules is less than the force of their interaction with the molecules of the solid. The force of their interaction with the molecules of the solid (as well as the force of gravity) makes the liquid spread over the surface of the solid (see Figure 2.21b): the shape of the liquid drop results from the action of the gravity force and the force of interaction of the liquid's molecules and the solid's molecules. The measure of wetting is the angle θ between the surface being wetted and the tangent to the surface of the liquid. This angle is called a contact angle or a wetting angle.

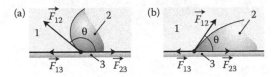

FIGURE 2.21 Schematic showing a drop balanced on the surface of a solid in the case of nonwetting (a) and wetting (b) liquids: (1) gas, (2) liquid, and (3) solid.

Many works have been concerned with the methods for determining the cleanliness of solids from their wettability with liquid drops [118,134–137]. But these methods suffer from the drawback that they evaluate the level of surface cleanliness by the state of a static drop, which is a time-consuming approach.

The physicochemical parameters controlling the thermodynamic wettability of solid surfaces have been clarified through the efforts of Zisman and others [138–140], while the understanding of deviations from the thermodynamic equilibrium is still in its initial stage, especially for the liquid–solid interface.

2.8 EVALUATING THE CLEANLINESS OF A SUBSTRATE FROM THE DYNAMIC STATE OF A LIQUID DROP DEPOSITED ON ITS SURFACE

References 141 and 142 propose a method and a device for evaluating cleanliness from the rate at which a liquid drop spreads over the test surface. The method is based on capturing the light flux formed by a drop (a liquid lens). But further studies have shown that this method is suitable for evaluating the cleanliness level of only a relatively contaminated substrate (a liquid drop deposited on an optically smooth surface fine-cleaned in a plasma-chemical etcher ceases to be a lens because of the substrate's good wettability). Patent [143], however, proposes a method for evaluating the level of substrate cleanliness from the dynamic state of a liquid drop deposited on a surface, including surfaces after fine cleaning [20,132]. The analyzer used in the method comprises a table, a VS-FAST/CG6 high-speed video camera, a sample-lighting system, and a PC for analyzing images (see Figures 2.22 and 2.23).

The analyzer functions as follows. The surface of the horizontally placed test substrate (5) is lit by a uniform light flux from the light source (1) through an infrared filter required to prevent the substrate surface from heating. A liquid drop (6) of a given volume is deposited on the substrate surface with the dispenser (4). The high-speed

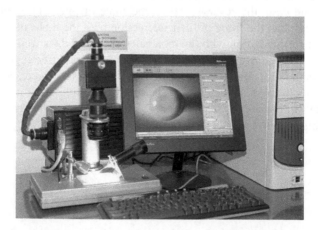

FIGURE 2.22 External view of the automated micro- and nanoroughness analyzer for substrate surfaces.

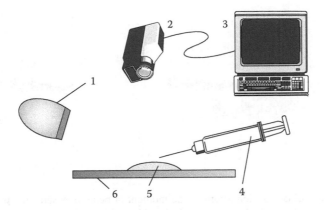

FIGURE 2.23 Components of the analyzer.

video camera (2) positioned perpendicularly to the substrate surface and focused on the area of interest records the spreading of the liquid drop.

The time during which the drop spreads over the test surface should not exceed 6 s (the time is limited by the camera's memory capacity). The image recorded by the recorder (3) in single-frame advance mode allows the parameters of the drop's dynamic state to be viewed and measured.

2.8.1 Description of the Experimental Method

2.8.1.1 Sample Preparation

ST-50-1 and Polikor substrates and graphite and glass photographic plates without emulsion layers were used as test samples.

The test substrates were cleaned with a standard technique consisting of the following steps: degreasing in a hot peroxide-ammonia solution (at 75–80°C); rinsing in running deionized water (to remove reaction products left from the preceding treatment); steeping in a hot sulfuric acid solution (at 90–100°C); rinsing in running deionized water (to remove residual acids) [144]; and etching in a UPT PDE-125-008 plasma-chemical etcher with a gas mixture containing 80% nitrogen and 20% oxygen at a pressure of $2.1 \cdot 10^{-1}$ Pa and a generator power of 500 W for a period of 4 min.

2.8.2 Description of the Experimental Procedure

A drop of distilled water was dispensed onto the surface of the prepared substrate immediately after cleaning. The falling and spreading of the drop were recorded with the VS-FAST/CG6 video camera at a rate of 1,000 frames per second.

The fall and spread process falls into three stages. The first stage consists of a drop forming at the tip of the dispenser needle (Figure 2.24); the second, in the drop falling by gravity onto the substrate surface (see Figure 2.25); and the third, in the drop undergoing a decaying oscillatory motion in spreading over the surface (see Figures 2.26 through 2.29).

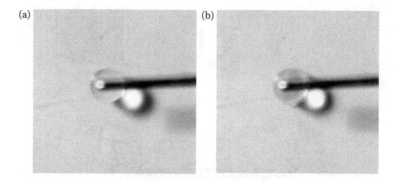

FIGURE 2.24 Initial position (drops at the moment of being released from the tip): (a) an ST-50-1 substrate cleaned by plasma-chemical etching with an air-exposure time of less than 1 min and (b) an ST-50-1 substrate cleaned by plasma-chemical etching with an air-exposure time of 20 min.

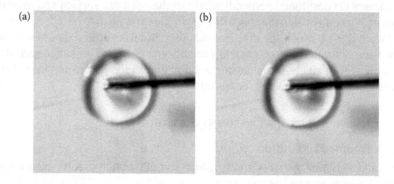

FIGURE 2.25 Drops in contact with the substrate surface: (a) an ST-50-1 substrate cleaned by plasma-chemical etching with an air-exposure time of less than 1 min and (b) an ST-50-1 substrate cleaned by plasma-chemical etching with an air-exposure time of 20 min.

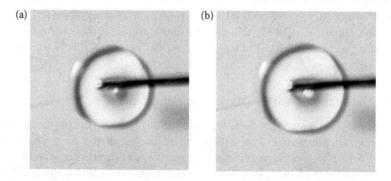

FIGURE 2.26 State of the drops captured 1 ms after the drops fell from the initial position: (a) an ST-50-1 substrate cleaned by plasma-chemical etching with an air-exposure time of less than 1 min and (b) an ST-50-1 substrate cleaned by plasma-chemical etching with an air-exposure time of 20 min.

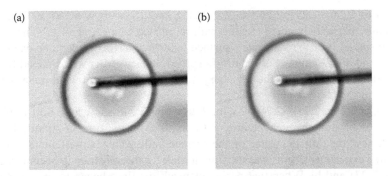

FIGURE 2.27   State of the drops captured 4 ms after the drops fell from the initial position: (a) an ST-50-1 substrate cleaned by plasma-chemical etching with an air-exposure time of less than 1 min and (b) an ST-50-1 substrate cleaned by plasma-chemical etching with an air-exposure time of 20 min.

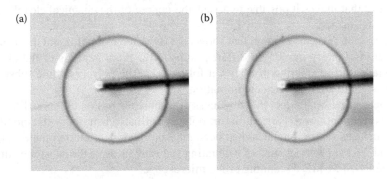

FIGURE 2.28   State of the drops captured 8 ms after the drops fell from the initial position: (a) an ST-50-1 substrate cleaned by plasma-chemical etching with an air-exposure time of less than 1 min and (b) an ST-50-1 substrate cleaned by plasma-chemical etching with an air-exposure time of 20 min.

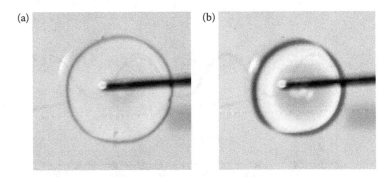

FIGURE 2.29   Static state of the drops captured 12 ms (a) and 65 ms (b) after the drops fell from the initial position: (a) an ST-50-1 substrate cleaned by plasma-chemical etching with an air-exposure time of less than 1 min and (b) an ST-50-1 substrate cleaned by plasma-chemical etching with an air-exposure time of 20 min.

Figures 2.24a, 2.25a, 2.26a, 2.27a, 2.28a, and 2.29a show an ST-50-1 substrate cleaned by plasma-chemical etching with an air-exposure time of less than 1 min. A drop spread over the substrate's surface in 12 ms, having undergone one oscillation. At the initial moment of the drop's spread, its wavefront was directed from the center radially outward to the periphery and the next moment, vice versa, with the drop taking a classic shape.

Figures 2.24b, 2.25b, 2.26b, 2.27b, 2.28b, and 2.29b show an ST-50-1 substrate cleaned by plasma-chemical etching with an air-exposure time of 20 min. Before it came to rest, the drop deposited on the surface underwent six oscillations in 65 ms.

In this experiment, 4 ms was the time it took for the substrate to become wet (see Figure 2.27a and b). When that period elapsed, the drop began to spread. And, as Figures 2.27b and 2.28b show, in the case of the substrate that had been exposed to the air for a period of 20 min after plasma-chemical cleaning, the drop's diameter remained virtually the same and comparable with that of the drop at rest (see Figure 2.29b). This means that the diameter of a spreading drop is a characteristic parameter that makes it possible to assess the degree of the chemical cleanliness of a substrate surface.

Figure 2.30 presents a graph that shows the dynamic state of liquid drops undergoing an oscillatory motion on different substrates. From analysis of the observations made during a period of 80 ms, it follows that the polycrystalline substrate is less resistant to atmospheric contamination.

To determine what effect the roughness of a substrate has on the spreading liquid drop, experiments were carried out on polycrystalline substrates with smooth and rough surfaces whose roughness had been measured with an interferometer. Figures 2.31 and 2.32 show the results of evaluating the roughness of the substrate surfaces with the P4-SPM-MDT scanning probe microscope.

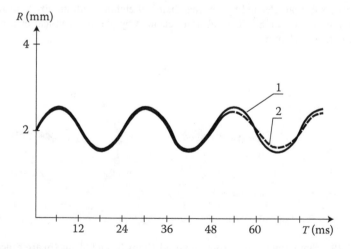

FIGURE 2.30 Dynamic state of distilled water drops on substrate surfaces that had been exposed to the air for a period of more than 1 month after plasma-chemical cleaning: (1) oscillations on a graphite substrate and (2) oscillations on a polycrystalline substrate.

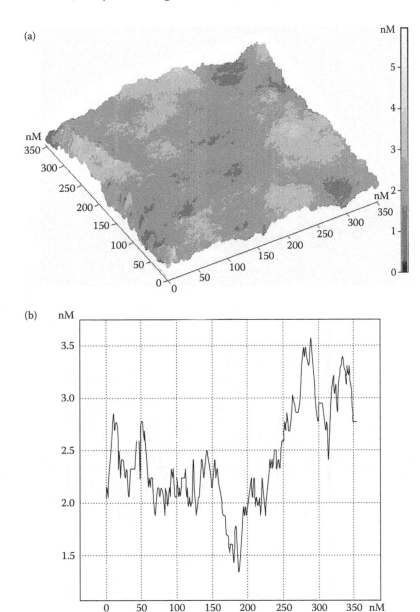

FIGURE 2.31 Measurement results for the surface of the ST-50-1 substrate: (a) structure and (b) profilogram.

The spreading of distilled water over the rough surface was similar to that on the smooth surface with the difference that the spreading time increased from 12 ms to 16 ms. The roughness parameters and microrelief shape observed make it possible to approximately determine an increase in the surface area of a drop's base: $S_{rgh} \approx 2S_{smth}$ of the same radius. The results of evaluating the activation energy of one molecule

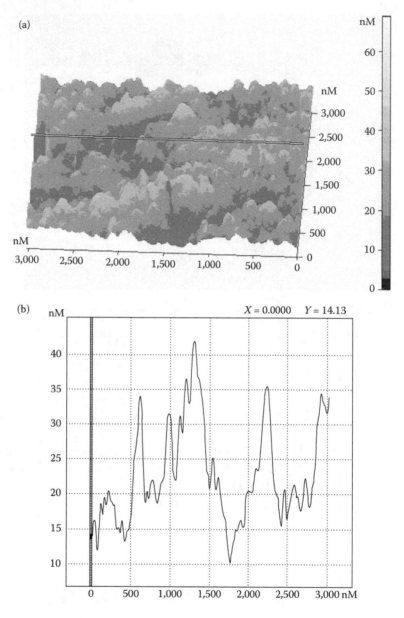

FIGURE 2.32 Measurement results for the surface of the ST-50-1 substrate: (a) structure and (b) profilogram.

for the smooth and the rough surface match with an accuracy of up to 90% for the same substrate material.

To determine what effect the roughness of a substrate has on the spreading liquid drop, experiments were carried out on the identically cleaned surfaces of glass plates with a roughness of 5–80 nm. Surface roughness was determined with a scanning

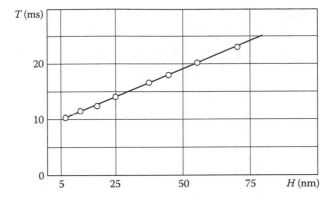

FIGURE 2.33 Relationship between surface roughness and the time that a distilled water drop takes to spread over the glass substrate of a VRP plate.

probe microscope by directly measuring the height from the bottom of the deepest depression to the peak of the highest protrusion in the three-dimensional (3D) image of the surface area of interest. In Reference 132, the area measures 350×350 nm. Figure 2.33 shows the relationship between surface roughness and the time that a distilled water drop takes to spread over the glass substrate of a VRP plate. In the range 6–70 nm, the relationship is linear.

The maximum spread radius of the liquid drop on the rough surface is less than that on the smooth surface by 2%–5%. With a roughness of more than 50 nm, a decrease in the spreading rate causes a reflected wavefront to appear on the surface of the spreading drop.

Therefore, the criterion of surface cleanliness is the maximum spread radius, and the criterion of surface smoothness is the minimum spreading time.

2.9 SPECIFICATIONS OF THE MICRO- AND NANOROUGHNESS ANALYZER

The specifications are as follows:

- Types of test substrates: VRP glass photographic plates; ST-50 substrates, Polikor substrates, and others
- Measurement range for organic contaminants: 10^{-7}–10^{-9} g/cm^2
- Measurement range for contaminant roughness: 5–200 nm
- Thickness of test substrates: 0.5–3.0 mm
- Liquid used: water
- Time required for evaluating the state of a substrate surface: not exceeding 1 min

The analyzer was used to optimize the cleaning regimes for VRP glass photographic plates measuring 50×50 mm. The plates were cleaned in the UPT PDE-125-008 plasma-chemical etcher with a gas mixture containing 80% nitrogen and

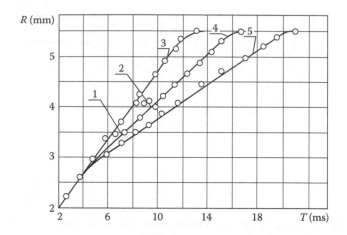

FIGURE 2.34 Relationship between the radius of the drop's base and the spreading time: (1) cleaning time = 30 s, surface roughness $N = 6$ nm; (2) cleaning time = 2 min, surface roughness $N = 6$ nm; (3) cleaning time = 3 min, surface roughness $N = 6$ nm; (4) cleaning time = 4 min, surface roughness $N = 40$ nm; and (5) cleaning time = 6 min, surface roughness $N = 60$ nm.

20% oxygen, at a pressure of $2.1 \cdot 10^{-1}$ Pa and a generator power of 500 W. The processing time for substrates from a single batch varied from 30 s to 6 min. Figure 2.34 shows the relationship between the spread radius of distilled water drops over the surfaces of VRP glass plates and the spreading time for different plasma-chemical treatment times.

The curves show that in the segment $T < 14$ ms, the spread radius increases. This means that plasma-chemical etching cleans the substrate surface but does not change its roughness. At $T = 14$ ms and a cleaning time of about 3 min, the spread radius reaches its maximum value while the roughness remains unchanged. When the cleaning time exceeds 3 min, the spread radius decreases insignificantly, while the spreading time continues to increase. This indicates an increased surface roughness. Therefore, the optimal cleaning time for this batch is 3 min.

The surface roughness of the substrates was measured with the P4-SPM-MDT scanning probe microscope. Figure 2.35 shows the surface of a substrate cleaned under conditions that are optimal for this type of substrate and this level of contamination. Its roughness does not exceed 6 nm.

Figure 2.36 shows the surface of a substrate cleaned under conditions that led to the pitting of the surface. Its roughness is about 60 nm.

This method makes it possible to quickly and efficiently evaluate the degree of surface cleanliness from the diameter of a spreading drop and evaluate nano-roughness from the time the drop takes to spread over the surface. From the dynamic state of a liquid drop deposited on a substrate, one can evaluate the efficiency of surface cleaning and quickly adjust the conditions of the cleaning process.

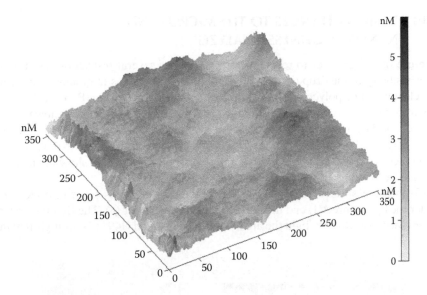

FIGURE 2.35 Three-dimensional image showing the surface of a glass substrate cleaned under optimal cleaning conditions.

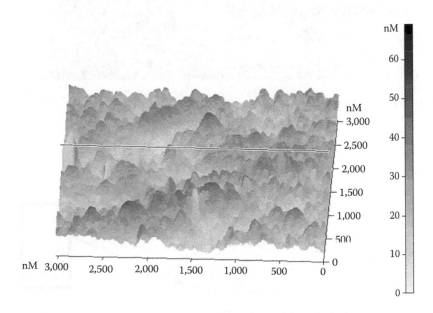

FIGURE 2.36 Three-dimensional image showing the surface of a glass substrate cleaned under sub-optimal cleaning conditions.

2.10 DESIGN CHANGES TO THE MICRO- AND NANOROUGHNESS ANALYZER

The use of the analyzer to register drop images has several restrictions relating to poor visibility of the drop (in particular, this causes problems in quickly measuring the cleanliness of polycrystalline and chrome-coated substrates). Poor visibility of the drop leads to errors in determining its shape and thus makes it impossible to draw correct conclusions regarding surface roughness and the presence of contaminants.

To improve the sensitivity and accuracy of measurements taken with the analyzer, References 145–148 propose changes to it. Figures 2.37 and 2.38 show an analyzer that has been modified with those changes in mind.

To measure surface roughness, the optical substrate (7) is placed horizontally under the light source (1) and the needle (5) of the dispenser (4). The lighting system is designed to uniformly light the area of the drop's spread [149–151] or highlight the

FIGURE 2.37 External view of the modified analyzer.

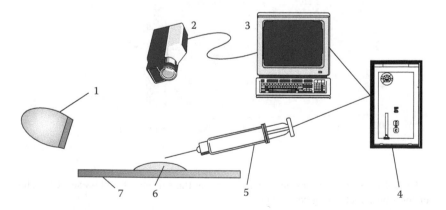

FIGURE 2.38 Components of the modified analyzer.

drop's shape with line-shaped lighting [152–155]. The operator uses the dispenser to deposit a liquid drop (6) on the substrate and simultaneously activates from the computer (3) the high-speed video camera (2) to record the drop's spreading over the surface. The recorded frames are then analyzed with image-processing algorithms and computer vision methods, and the size and shape of the drop, the rate at which it spreads in various directions, and its other dynamic characteristics are assessed [146–147]. Then the measurement results are compared with those for the reference substrates whose data are stored in the analyzer's database. With the comparison results in mind, conclusions are drawn regarding the degree of roughness and cleanliness and the presence of any defects, contaminants, and other irregularities on the test substrate's surface.

For measurement results to be reliable and repeatable, it is essential that the dispenser dose the liquid with high precision. The nuances of recording with the VS-FAST/CG6 camera should also be kept in mind: all recorded video data are first saved to its internal memory and then, as the memory fills up, transferred to the computer's RAM. As the rate at which data are transferred to the computer is less than the rate at which data are recorded to the camera's internal memory, the recoding start time needs to be synchronized with the moment the drop is deposited on the substrate. Otherwise the camera's internal memory will fill up before the spreading of the drop has time to be fully recorded. Since the known machine vision systems [156,157] are free from this problem, additional research is required to eliminate it in the analyzer. One possible solution is to use an automatic dispenser.

2.10.1 REQUIREMENTS FOR THE AUTOMATIC DISPENSER

An analysis found that the dispenser should:

1. Dispense liquid by command from the computer
2. Provide the capability to set the dosing rate and volume
3. Have a small dosing error
4. Include a software development kit
5. Be connectable to a PC via USB or LAN ports

With a manual dispenser, surface roughness was measured as follows: The test substrate was placed on the sample stage, and the high-speed video camera was switched to the recording mode. Then the operator manually deposited a liquid drop on the substrate with a mechanical dispenser. Because of having to manually intervene in the measurement procedure, the time between the start of recording and the moment that the drop falls onto the surface is a function of the operator's reaction time and varies in the range 0.3–1.2 s. With some of the computer time spent processing noninformative frames recorded during that interval, measurement times increase.

After a review of the specialized-equipment market, we chose Microlab ML630 Single Syringe PC Controlled, an electronic 50 µL syringe pump [158] from Hamilton, a US–Swiss company. With this syringe pump, liquid drops with a volume from 0.01 µL can be formed with an accuracy of 1%.

Microlab ML630 communicates with the PC over the 100Base-T (Fast Ethernet) network through a Cat 5 UTP cable attached to an 8P8C connector. Communication with the dispenser is accomplished by exchanging data over the TCP/IP protocol. This functionality is implemented in the manufacturer-supplied dynamic-link library.

2.10.2 COMPATIBILITY OF THE PUMP'S CONTROL MODULE AND THE ANALYZER'S SOFTWARE

Integrating the syringe pump with the hardware and software of the micro- and nanoroughness analyzer poses the problem of compatibility. The analyzer's control software was written in C++ without using the Microsoft .NET library, while the syringe pump's control library was written in C# using the Microsoft .NET library. To make it possible to call the syringe pump's control functions from the code written in C++, the Unmanaged Exports library [159] was used to create a module for translating C++ calls into .NET calls and back.

Some difficulty was encountered with the nuances of operating the functions from the syringe pump's control library. To form a drop, the dispense function is called, which receives information on the volume to be dispensed as an input parameter. When this function is called, the syringe pump receives a signal that activates the electrical drive of the syringe piston. After the piston has moved to the required position, the drive slows down and stops. A return from the dispense function to the main program can be made only after the piston has stopped moving and the microprogram of the drive controller has scanned the hardware for faults. By this time the dispensing process will have finished: the drop will have reached the substrate surface and stopped spreading. Therefore, a recording of the drop spreading on the surface has to start before the dispense function is called.

But in this case, extraneous frames will be recorded that show the surface before the drop has fallen onto it. Experiments have shown that a drop reaches the substrate 480–580 ms after the dispense function is called. This means that during recording at 500 frames per second, the first 240–290 frames will be noninformative and have to be skipped, thereby leading to inefficient use of the camera's internal memory. To sidestep this problem, the activation of the dispensing process and the recording process was divided into individual execution threads: The calling of the dispense function was assigned to the alternative thread and the recording process was configured to start in the main thread with a delay of 480 ms.

Some difficulty was caused by having to take into account delays due to the time needed to refill the syringe. Before measurement starts, the syringe pump goes through an initialization procedure during which the residual liquid is removed from the syringe and a new portion added. After forming several drops during several consecutive measurements, the syringe is emptied and needs to be refilled before the next measurement. Experiments have shown that the time between a call of the dispense function and the moment that the drop touches the surface of the test substrate depends on whether that call is the first one after the syringe is refilled. For instance, if the dispense function is called for the first time after the syringe is refilled, the time between the call and the moment the drop touches the surface is 380–450 ms as opposed to 480–580 ms for the second and subsequent calls. The difference is due to the nuances of how the syringe pump's internal components function.

As a result, the software was supplemented with a function for recognizing such situations and for setting, if necessary, an appropriate delay before the recording starts (depending on whether the syringe needs to be refilled before the next measurement).

2.10.3 OVERVIEW OF THE OPERATION OF THE MODIFIED ANALYZER

Figure 2.39 presents a graph showing the periods between the start of recording and the moment that the drop touches the substrate surface for the manual and the automatic dispensing modes. For recording at a rate of 500 frames per second, the average number of noninformative frames (frames recorded before the drop touched the surface) for 30 experiments diminished from 276 for manual dispensing to 36 for automatic dispensing.

Replacing the mechanical dispenser with the PC-controlled Microlab ML630 syringe pump helped improve the accuracy of forming liquid drops with the desired volume. Thanks to the larger number of fixed piston positions and smaller distances between them, the automated syringe pump provides the capability to set up the measurement conditions more accurately for taking, under various ambient conditions, measurements on substrates made of various materials.

Equipped with an electronic dispenser, the micro- and nanoroughness analyzer is an important step toward fully automated measurements. Eliminating the error inherent in the manual dispensing method avoids the loss relating to the processing of noninformative frames. This offers an added advantage of reducing overall measurement times.

Automating the horizontal movement of the test-substrate base will make it possible to evaluate the roughness and cleanliness of the entire substrate surface without manual intervention.

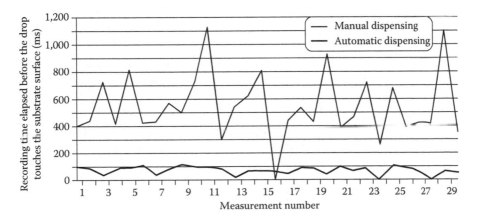

FIGURE 2.39 Graph showing the recording times of noninformative frames for manual and automatic dispensing.

2.11   CHAPTER SUMMARY

The nondestructive tribometric method and the device for quickly measuring the cleanliness of substrate surfaces discussed in this chapter do not require the use of reference surfaces for calibration purposes or probes whose surfaces require special treatment. With the proposed method, the coefficient of sliding friction is used to measure surface cleanliness. This extends the range of measurable levels of substrate cleanliness to 10^{-6}–10^{-10} g/cm^2 (which is broader by comparison with Russian and foreign equivalent methods) and reduces the measurement time to 5–15 s.

The proposed changes to the tribometer—a special optical grating incorporated into its optical system and the automated processing of experimental data—have dramatically improved its resolving power.

With the measurement method based on evaluating the dynamic state of the liquid drop, the degree of surface cleanliness can be quickly and efficiently evaluated from the diameter of a spreading drop, and its spreading time allows nanoroughness to be measured. A substrate surface cleaned by plasma-chemical etching can be evaluated with a high-speed video camera. From the dynamic state of a liquid drop deposited on the substrate surface, we can evaluate substrate-cleaning practices and how efficient they are for cleaning the nanoroughness of substrate surfaces.

3 Increasing the Degree of Surface Cleanliness with Low-Temperature Off-Electrode Plasma

A major problem hindering the development of diffractive optics and micro- and nanoelectronics is how to fabricate surfaces with the desired degree of cleanliness. Residual resist layers as well as organic molecular compounds (solvents, chemicals, etc.) adsorbed at wafer surfaces are the primary sources of contamination [160,161]. This necessitates cleaning organic contaminants off substrate surfaces before etching.

Today, the most common surface-cleaning techniques are chemical cleaning, laser cleaning, and plasma cleaning. To generate uniform plasma, plasma techniques use high-frequency (HF) and superhigh-frequency (SHF) sources that are complex, costly, and energy-intensive. Treating wafers with plasma generated by such sources contaminates surfaces with low-active or inactive particles because the substrate is placed between the electrodes of the gas-discharge device. Plasma parameters in this case are determined chiefly by the properties of the surface being treated (the loading effect).

Chapter 1 demonstrated the advantages of using off-electrode plasma. In that technique, ion-plasma fluxes are formed outside the electrodes, and the possibility of the surface becoming contaminated with inactive plasma particles is ruled out because only negatively charged particles (ions and electrons) move toward the surface. But because contemporary literature is silent on theoretical and experimental research into the mechanism of surface cleaning that uses off-electrode plasma, no practical methods are yet available.

This chapter analyses widely used conventional cleaning techniques and theoretically and experimentally investigates the mechanism of surface cleaning with directed fluxes of low-temperature plasma generated by a high-voltage gas discharge outside the electrode gap. The discussion is aimed at creating efficient cleaning methods to improve the fabrication quality of optical microreliefs.

3.1 OVERVIEW OF METHODS FOR SURFACE CLEANING

The major difficulty in forming a mask during the fabrication of, for example, a diffraction microrelief on large-format substrates is to obtain a technologically clean surface with homogeneously and uniformly distributed properties. This calls for higher requirements for cleaning various contaminants off substrate surfaces [66,108,162–165].

Existing technology for obtaining technologically clean surfaces offers techniques for rough and final cleaning. Rough cleaning serves to remove primary macrocontaminants and uses chemical etching with acids, alkalis, and solutions. But even after a substrate has been thoroughly rinsed in running distilled or deionized water, its surface will invariably be contaminated with the residual etching reagent. To remove this type of contamination, final-cleaning techniques are used that involve the use of highly pure chemicals or surface bombardment with low-temperature plasma particles. Final cleaning has a decisive effect on mask adhesion and takes place immediately before a mask is deposited on the substrate surface.

This section discusses the existing final-cleaning techniques widely used in diffractive optics and microelectronics.

3.1.1 CHEMICAL CLEANING

Chemical cleaning consists of removing various types of contaminants from substrate surfaces through various methods: immersion in a washing solution or the saturated vapor of a liquid; boiling; ultrasonic treatment with complexones and surface-active agents (SAAs); hydromechanical treatment; rinsing in distilled and deionized water; and so on [109,166–175]. All these methods involve using many highly pure chemicals [168,174], process tooling made of chemically resistant materials, and various energy sources. This significantly complicates the environmental conditions of techniques used.

An example of the complexity inherent in chemical cleaning is the Lada-150 wafer-cleaning line comprising up to 10 process modules. Substrate treatment in such equipment takes place in a nitrogen-filled closed manufacturing system in which wafers are not exposed to the air [172]. Highly toxic chemicals such as $H_2SO_4 + H_2O_2$, $NH_4OH + H_2O_2 + H_2O$, $HF + H_2O$, $HCl + H_2O_2 + H_2O$, and $Na_2Cr_2O_7 + H_2SO_4 + H_2O$ solutions [174,175] are used as processing media. The trend to use SAAs to reduce toxicity does not solve the problem in its entirety since any chemical in large quantities is a source of environmental pollution—primarily, waste water pollution [109]. And since chemical-cleaning process flows are, as a rule, designed for specific types of substrates and contaminants, the entire process flow has to be reworked each time a parameter is changed. This accounts for the large number of works on new methods and techniques for chemical cleaning of substrate surfaces [164–173,175,176].

Thus, the chemical cleaning techniques have the following drawbacks:

- Necessity of using highly pure chemical reagents.
- High toxicity affecting the operability of equipment and requiring special measures to dispose of waste to reduce environmental pollution.
- Necessity of selecting an appropriate solvent (acid, alkali, SAA) for each type of contaminant.
- A broad range of process equipment and tooling (degreasing equipment, ultrasonic baths, vibrating centrifuges, drying ovens, exhaust hoods, etc.)
- Surface contamination with both residual reagents and the impurities contained in them [109].

But in the absence of versatile, efficient final-cleaning techniques, chemical techniques continue to be widely used in diffractive optics and microelectronics [109,128].

3.1.2 LASER CLEANING

Laser cleaning relates to "dry" treatment methods and consists of removing various types of contaminants from the surface by heating surface contaminants to a temperature sufficient for thermal desorption. The radiation intensity required to remove a given type of contaminant is selected from the range $P \approx 10^3$–10^4 W/cm^2— that is, the intensity values are relatively low. Surface contaminants are removable by using the surface-melting method. Even with the radiation intensity increased to 10^4–10^5 W/cm^2, the method does not destroy the substrate material. But when the liquid-phase crystallization of the surface melt takes place, impurity atoms may spread into the depth of the molten layer, thereby adversely affecting the substrate properties. Impurity atoms are observed in the composition of surface contaminants and in the working chamber's atmosphere and spread according to their segregation coefficient.

A major drawback of laser cleaning is that radiation intensity is highly nonuniform across the beam [109,177–179]. This drawback results in nonuniform surface cleaning when an automated substrate-scanning system is used and in the possibility of mechanical stress occurring on the substrate surface that may cause the microrelief to deform during subsequent etching.

Because of the broad range of substrate materials used, surface cleaning requires the use of different lasers with a wavelength ranging from the mid-infrared to the deep ultraviolet. References 109, 178, and 179 discuss industrial laser-beam generators: CO_2 and CO lasers, yttrium-aluminum-garnet ($Y_3Al_5O_{12}$) lasers, ruby ($Al_{2-x}Cr_xO_3$) lasers, ultraviolet exciplex lasers, and copper vapor lasers. But none of these lasers is yet capable of efficiently cleaning all types of substrates [109,180]. As a result, many works have been devoted to modifying laser systems and methods for their use [109,177–180] in order to obtain technologically clean surfaces.

To summarize, laser technology for surface cleaning has the following drawbacks:

- Nonuniform cleaning and the necessity of using scanning systems that make the equipment more complex and increase its cost
- Surface-layer defects such as craters and wavelike relief

Because of these drawbacks, lasers are suitable only for highly specialized applications.

3.1.3 LOW-TEMPERATURE PLASMA CLEANING

Cleaning of substrate surfaces with low-temperature plasma takes place under vacuum conditions before film coatings are applied [181–186]. In this case, cleaning efficiency depends on the type of discharge used, the gas-discharge device's design features, and substrate size. For instance, when a substrate is cleaned with a glow discharge, the substrate is placed on the anode's working surface. As a result, the gas-discharge device's parameters become dependent on the surface area being treated—that is, a loading effect [109,128,187] occurs.

Cleaning of large surface areas with magnetron-discharge plasma requires scanning the surface, because the charged particle flux generated by this discharge is nonuniform in its cross section. And in this case the surface is generally cleaned with either positive Ar ions (i.e., inert particles) or O ions. Because both processes involve physical sputtering [188], an increase in the efficiency of surface cleaning entails increasing the charged particles' energy, and this deteriorates the surface properties [109,128].

A major drawback with HF and SHF discharges is the loading effect, which is observed as a change in the density of the particles colliding with the surface in response to a change in the number of substrates or their surface areas. To stabilize gas-discharge parameters, optimal resonant frequency, power, gas inleakage rate, and other parameters must be determined. For instance, in the case of modern equipment that uses transformer-coupled plasma and capacitively coupled plasma, the presence of a drawing potential necessitates increasing the dimensions of the plasma-forming device by several times for larger substrates [184–186].

Another drawback with HF and SHF discharges is that their relatively high power (300–600 W) causes the surface properties of substrates to deteriorate [184–186,189]. To limit the deterioration, chemically active gases and gas mixtures are used that determine cleaning efficiency. Gases used must yield a large number of reactive particles that form volatile compounds with surface atoms, be nontoxic and nonexplosive in the gas phase and the gas-vapor phase, and noncorrosive. And the gases must not contaminate intrachamber components and evacuation lines or have a highly adverse effect on vacuum-pump parts and oil if vacuum pumps are used.

The most common techniques for low-temperature plasma cleaning involve using highly pure gases and gas mixtures (H_2, O_2 + Ar + H_2 + N_2 + He, C_xF_y, CF_4 + O_2, Cl_2, HCl, NF_3, SF_4, etc.). This significantly increases the cost of fabricating microelectronic and diffractive optical elements [34,35,109,128,162,177,181–185,187,189–194]. Abandoning the expensive gas techniques in favor of in-the-air cleaning would have a significantly beneficial economic effect.

Thus, surface cleaning with a low-temperature plasma flux generated with the devices outlined above has serious drawbacks whose elimination complicates the equipment, tooling, and operating conditions.

3.2 FORMATION MECHANISMS OF SURFACE PROPERTIES

As the properties of a substrate surface depend on the concentration of atoms and molecules adsorbed at the surface, surfaces can be classified into two categories: atomically clean surfaces and technologically clean surfaces.

An atomically clean surface is a surface that is free from foreign atoms and that is obtained under ultrahigh vacuum conditions through ion bombardment, high-temperature annealing, monocrystal cleaving, and other techniques [109,128,190]. For an atomically clean surface, the microrelief etch rate does not vary across the entire mask–substrate interface, and etching does not distort microstructure shapes.

A technologically clean surface is a surface with certain physicochemical properties that contains adsorbate atoms in a quantity that affects neither the operating characteristics of the component being fabricated on that surface nor any subsequent processing.

Analyzing the composition of surface contaminants allows the contaminants to be classified into three categories:

1. Contaminants adsorbed at the surface from the environment
2. Contaminants from contact with process tooling, such as residual acids, alkalis, solutions, and other process materials
3. Contaminants adsorbed at the surface directly in the working chamber before etching, such as organic molecules' atoms present even in the high-vacuum environment of a dry-pumping system or vacuum-oil vapors from the forevacuum system in UVN vacuum setups [130,187,195]

Some atoms and molecules in categories 1 and 2 are capable of chemisorption—that is, of generating strong bonds at the surface. Currently, there are well-developed dry and wet etching techniques for removing contaminants [16,17,109,128,190,191,193,196–198]. For example, a standard chemical washing technique will remove most surface contaminants.

During final cleaning, removing some atoms and molecules in category 2 requires using expensive plasma-chemical equipment [16,17,34,35,187,199,200].

Category 3 contaminants are hardest to remove: they form at the substrate surface as separate atoms or monomolecular layers directly during the evacuation of the working chamber. In terms of how difficult contaminants are to remove, of the greatest interest is the UVN-2M-1 vacuum evaporator, which incorporates oil-diffusion and forevacuum pumps and which is widely used in the microelectronic industry. Process-liquid vapors constitute the primary contamination source. When a mineral oil is used (such as VM-1), their flows reach the level of $0.5\ \mu g/h \cdot cm^2$ [130,187]. That is why, for purposes of this monograph, VM-1 vacuum oil, whose properties are well known, was used as an organic-contaminant simulator. To determine the mechanism of how molecules are distributed in such flows during the evacuation of the working chamber, the test substrates were placed on a substrate carousel (positions I, II, III, and IV in Figure 3.1).

The adsorption rate of vacuum-oil molecules at the substrate surfaces was controlled by changing the time of evacuation with a mechanical pump (to a pressure of 3.5 Pa) and a diffusion pump (to a pressure of $3.99 \cdot 10^{-3}$ Pa). Figure 3.2 shows the results of the experiments. Surface cleanliness was measured with the technique presented in Reference 70.

As curves 1, 2, 3, and 4 show, the substrates with chemically cleaned surfaces are characterized by a significantly increased degree of contamination in the ranges $T < 5$ min (curves 1 and 2), $T < 5.4$ min (curve 3), and $T < 7$ min (curve 4). The increase is presumably due to adsorption of atmospheric hydrocarbons present in the vacuum chamber and of vacuum-oil molecules brought by their backflow from

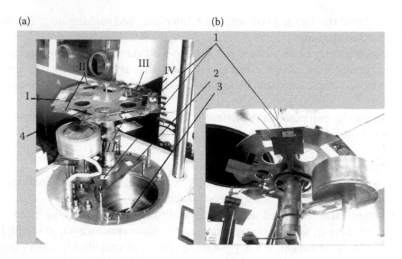

FIGURE 3.1 Arrangement of substrates in UVN-2M-1's working chamber: (a) top view and (b) bottom view. (1) Test substrate, (2) carousel base, (3) diffusion-pump valve, and (4) substrate carousel.

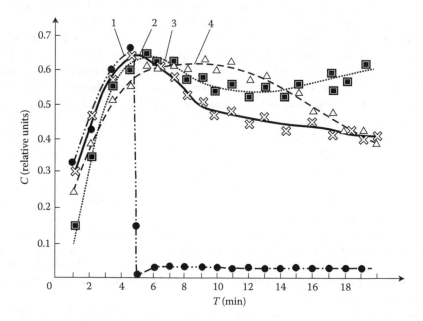

FIGURE 3.2 Curves characterizing the degree of surface contamination with vacuum-oil molecules: (1) curve characterizing surface cleanliness before ($T \leq 5$ min) and after ($T > 5$ min) final cleaning and (1)–(4) substrate contamination for different substrate positions on the carousel.

the mechanical and diffusion pumps. The drop of curves 2, 3, and 4 outside the ranges indicates desorption of these molecules from the surface.

As the flows of vacuum-oil molecules from the evacuation system are nonuniform, the degree of their effect on substrate contamination differs depending on the position of a given substrate on the carousel (positions I, II, III, and IV in Figure 3.1). For instance, during the 2-h evacuation period, the substrates placed on the carousel base (2)—that is, those in a region outside the direct impact of the oil backflow—were contaminated 1.6 times less than those placed on the carousel surface.

From this it follows that the primary contaminant in UVN-2M-1's working chamber is vacuum oil, a product used in its evacuation system.

The substrates in position II on the carousel, on reaching a contamination level corresponding to $T = 5$ min, were bombarded with off-electrode plasma particles at a discharge current of $I = 10$ mA and an electrode voltage of $U = 1.2$ kV for a period of $t = 10$ s (see curve 1 in Figure 3.2). The manner in which curve 1 changes shows that bombardment with plasma particles efficiently cleans vacuum-oil molecules off the substrate surface. At $T > 5$ min, the surface contamination increases slightly by no more than 3%.

With these results in mind, we conclude that:

1. Contamination of substrate surfaces in the vacuum system's working chamber is a function of how the backflow of vacuum oil from the vacuum-pump system is distributed.
2. Primary contamination takes place within the first five minutes of evacuation with a mechanical pump; then the atoms and molecules of organic contaminants are desorbed.
3. Efficient cleaning must take place as part of forming a forevacuum at $T \geq 5$ min.

3.3 MOLECULAR STRUCTURE ANALYSIS OF THE ORGANIC CONTAMINANT

A characteristic feature of the vacuum-oil molecule is its spatial arrangement on the substrate surface (see Figure 3.3) [201,202].

As the bond angles in the six-membered ring are equal to the normal valence angle of carbon (109° 28′), the bonds do not deviate from their normal direction and

(a)   (b)

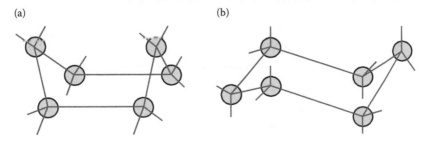

FIGURE 3.3   Carbon skeletons of cyclohexyl: (a) tub shape and (b) armchair shape.

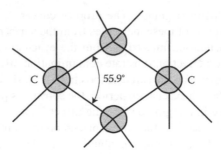

FIGURE 3.4 Carbon skeleton of cyclohexyl.

the cyclohexyl ring is highly stable. As a rule, cyclohexyl bonds have an armchair shape because the atomic binding energy in this molecule is less than that in a molecule with a tub-shaped carbon skeleton [201]. The cross section of the armchair-shaped cyclohexyl molecule is characterized by large height dimensions (Figure 3.4).

Consider the spatial arrangement of carbon and hydrogen atoms in the oil molecule whose main component is $C_{33}H_{64}$ [201]. For purposes of this discussion, we will divide this molecule into three characteristic parts (see Figure 3.5): $C_6H_{11}-$; $-(CH_2)_{25}-$; and $-CH = CH_2$. In the first part, the arrangement of carbon atoms and hydrogen atoms corresponds to that in the ethylene molecule [202], except that one hydrogen atom is replaced with a carbon atom (see Figure 3.5). The second part is a long side chain consisting of CH_2 atom groups.

Adsorption of oil molecules at substrate surfaces is determined by Van der Waals forces [99], and the resultant bond is comparatively weak. For instance, the heat of physical adsorption of vacuum-oil vapors on carbon is $90-100 \cdot 10^6$ J/kmol [201]. Depending on whether the adsorbed molecule and the adsorbent have polar properties, the following characteristic types of interaction can occur during physical adsorption [203]:

- Adsorption of nonpolar molecules (H_2, O_2, N_2, etc.) on a nonpolar adsorbent (solids with atomic and molecular lattices such as graphite, diamond, and organic-substance crystals).
- Adsorption of nonpolar molecules on a polar adsorbent (solids with ionic lattices).
- Adsorption of polar molecules on a nonpolar adsorbent.

$$
\begin{array}{c}
CH_2 \\
CH_2 \qquad\qquad CH - (CH_2)_{25} - CH = CH_2 \\
CH_2 \qquad\qquad CH_2 \\
CH_2
\end{array}
$$

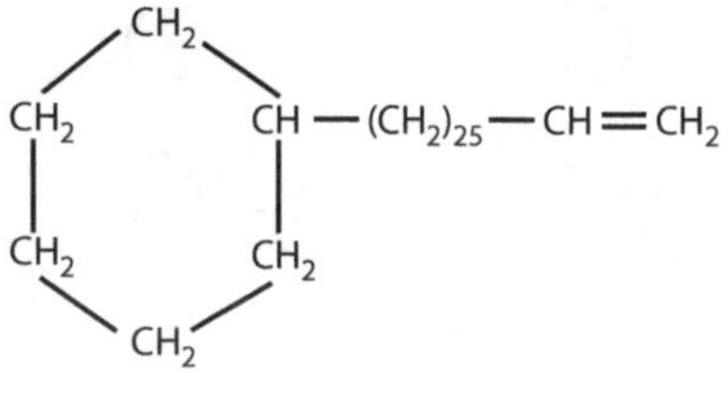

FIGURE 3.5 Structural formula of the vacuum-oil molecule.

The interaction mechanism of polar molecules and polar adsorbents is similar to the mechanism whereby a polar adsorbent adsorbs a nonpolar molecule [203]. Because vacuum oils are based on the component $C_{33}H_{64}$, all hydrocarbons are non-polar [201]. Therefore, the usual dispersion forces are supplemented by induction forces—the positive and negative ions of the solid's lattice that induce opposite-sign charges in a nonpolar molecule. This results in an additional electrostatic attraction of the adsorbed molecule by the polar adsorbent.

According to Reference 187, the binding energy between the oil-molecule atoms and the substrate atoms is greater than that between the oil-molecule atoms; we can suppose that even when oil molecules are effectively desorbed from the sub-strate surface, a monomolecular contaminant layer will invariably be present on it [109,187] and that the layer's thickness depends on the atoms' structural arrangement (see Figure 3.6).

The thickness of the oil-molecule layer can be derived from the expression

$$h_f = h_K + 2 \cdot h'_{CH} + \frac{h_H}{2} + h_{ad}, \tag{3.1}$$

where h_f is the film thickness (nm); h_K is the distance between opposite carbon atoms in the C_6H_{11} ring (nm); h'_{CH} is the projection of the C–H bond's length on the axis perpendicular to the substrate surface (nm); h_H is the orbital radius of the hydrogen atom (nm); and h_{ad} is the distance determined by film–substrate adhesion (nm).

The molecule structure in Figure 3.6a yields the following equation:

$$h_K = h_{CC} + 2 \cdot h_{CC} \cdot \sin\left(90 - \frac{\alpha}{2}\right), \tag{3.2}$$

where h_{CC} is the length of the C–C bond (nm).

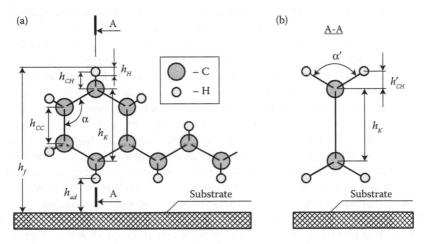

FIGURE 3.6 Arrangement of an oil molecule on the substrate surface: (a) longitudinal cross section and (b) transverse cross section.

The value of $h_{CH}^{\mid}$ can be determined from the transverse cross section shown in Figure 3.6b:

$$h_{CH}^{\mid} = h_{CH} \cdot \cos\left(\frac{\alpha'}{2}\right),\tag{3.3}$$

where h_{CH} is the length of the C–H bond (nm) and α' is the bond angle of CH_2 atoms.

It is known that $h_H = 0.05$ nm [202], $h_{CH} = 0.109$ nm, $h_{CC} = 0.154$ nm [202], and $h_{ad} = 0.5$ nm [203]. Substituting the numerical values into Equations 3.1 through 3.3 gives

$$h_f = 0.154 + 2 \cdot 0.154 \cdot \sin\left(90 - \frac{109.47}{2}\right)$$
$$+ 2 \cdot 0.109 \times \cos(58.35) + \frac{0.05}{2} + 0.5 = 0.974 \, \text{nm.}$$

Because the h_f obtained from this method agrees well with the data presented in Reference 187, we will assume, for purposes of further discussion, that the thickness of the oil film on the substrate surface equals $h_f \approx 1$ nm.

Thus, during evacuation, a hard-to-remove monomolecular layer of vacuum-oil molecules forms on substrate surfaces.

3.4 PREPARING INITIAL SAMPLES WITH A GIVEN DEGREE OF CONTAMINATION

The investigation was carried out on $30 \times 20 \times 1$ mm polycrystalline and silicon dioxide (SiO_2) substrates because these two types are the most widely used in fabricating various DOEs and nano- and microelectronic elements.

There are several methods for depositing monomolecular films of organic contaminants on the substrate surface: immersing substrates in a vacuum-oil solution and then in toluene or dichloroethane, followed by drying [18]; centrifuging a vacuum-oil solution; and spreading an oil drop on a water surface (the Langmuir–Blodgett method) [204,205]. But these methods do not allow a film structure to be formed whose qualitative composition would be identical to that of contaminants formed on a substrate in the vacuum chamber, because the solvents act as contaminants themselves and solvent molecules will be present at the substrate surface along with vacuum-oil molecules.

Further, vacuum-oil molecules are polydisperse—that is, each of them can have different molecular weights. This causes the film to be nonuniformly distributed on the substrate surface when the methods of immersion and centrifuging are used. The Langmuir–Blodgett method is also inapplicable: it gives densely compressed film structures that do not match those of films formed in the vacuum chamber.

To overcome these shortcomings, this section proposes using the vacuum evaporation method involving the use of the surface contaminator shown in Figure 3.7 for depositing monomolecular films whose properties and structure match those of substrate contaminants formed in the vacuum chamber.

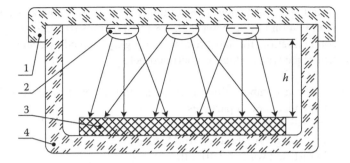

FIGURE 3.7 Design of the surface contaminator [136]: (1) quartz-vessel lid, (2) drops of VM-1 vacuum oil, (3) substrate, and (4) quartz vessel.

The contaminator is a closed quartz vessel. A substrate is placed into the quartz vessel, and then vacuum-oil drops (a source of organic contaminants whose properties are well known) are deposited on the lid's inner surface. The quartz vessel is covered with the lid and placed into UVN-2M-1's working chamber, which is then evacuated to a pressure of 1.33 Pa at a temperature of 300 K.

Under these conditions, the mean free path of evaporating oil molecules equals the distance between the drops' surfaces and the test substrate (≈5 mm). Because oil molecules reach the substrate surface without interacting with other compounds' molecules in the vacuum chamber, the contaminants adsorbed at the surface show uniform properties. Each contamination regime was reproduced at least 10 times, with the parameter spread no greater than 10%.

The initial substrates' surfaces had a large amount of contamination (see Figure 3.8a) whose numerical value, C_d, was measured in relative units with the technique described in References 70 and 71.

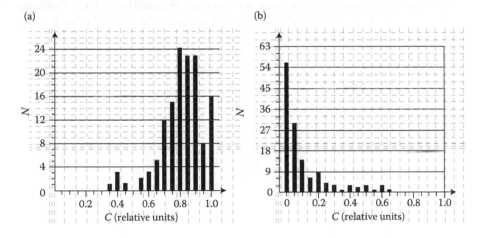

FIGURE 3.8 Graphs showing substrate distribution by degree of initial surface contamination: (a) contamination degree of initial substrates and (b) contamination degree of chemically cleaned substrates.

As the bars in Figure 3.8a show, 69% of the substrates (94 substrates out of 136) showed a contamination level greater than 0.8 relative units (0.802 · 10⁻⁷ g/cm²). To remove this contamination, the substrates were roughly cleaned by chemical etching through the technique whose steps are outlined below:

1. Rubbing with a cambric cloth soaked in ethyl alcohol on both sides, to remove large mechanical particles
2. Boiling in distilled water for 10 min
3. Boiling in an alkali solution for 10–15 min
4. Boiling in distilled water for 10 min
5. Boiling in ethyl alcohol for 10 min

This cleaning removed the greater part of contaminants with unknown parameters (see Figure 3.8b) from the substrate surfaces. But some substrates still showed a contamination level up to 0.65 relative units. This indicated the presence of hard-to-remove, chemically bonded contaminants formed at the time the substrates were made and removable only with special chemical or plasma-chemical cleaning techniques. As a result, all chemically cleaned substrates used in the experiment were screened to reduce the spread of cleanliness values to 0.03%. Then the substrates were placed on the bottom of the quartz vessel.

The concentration of contaminants at the test surface was controlled by changing the time (t) that the substrates were held in the contaminator inside the vacuum chamber. The value of t varied from 0 to 90 min. At $t = 60$ min and a chamber pressure of 1.33 Pa, the pulse time τ, which characterizes the degree of surface cleanliness, reduced to 0.05 · 10⁻² s (see Figure 3.9) and remained unchanged in the range $60 \leq t \leq 90$ min because of the saturated contamination process and the stabilized friction coefficient, which, according to Reference 109, are achieved when the concentration of organic contaminants at the surface reaches 10⁻⁷–5 · 10⁻⁶ g/cm².

An examination with an atomic force microscope confirmed that a film of vacuum oil forms on substrate surfaces subjected to contamination for a period of $t = 60$ min (see Figure 3.10).

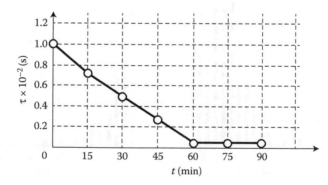

FIGURE 3.9 Relationship between the degree of contamination of substrate surfaces and the time that the substrates are held in the contaminator inside the vacuum chamber: $\alpha = 45°$ and $\beta = 5.5°$.

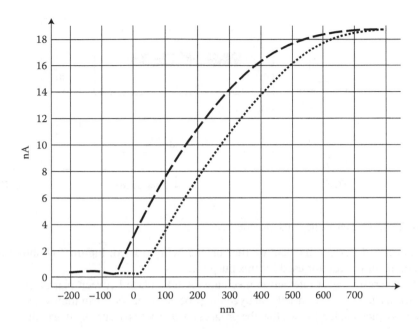

FIGURE 3.10 Force curve obtained with Solver PRO-M for the test surface before final cleaning. The interaction distance between the cantilever and the surface structure is a 3-nm film of organic contamination.

According to the calculations presented in Section 3.3, a 3-nm-thick film should comprise three layers of vacuum-oil molecules, but as reported in References 206–208, in examinations with an atomic force microscope, the molecular layer wets the cantilever, thereby causing hysteresis in measurements and increasing the related readings. When monomolecular films are measured with contact atomic force microscopy, vacuum-oil molecules may elongate and become polarized, which is an effect that determines the 3 nm thickness and agrees with the calculated molecule length. Calculating the height of the molecule (see Figure 3.6) turned to a vertical position yields a value of 3.41 nm. This signifies that the scanning probe microscope indeed measures the monomolecular layer of vacuum oil.

Thus, the initial sample surfaces were obtained by storing the substrates in the quartz vessel for 60 min at a pressure of 1.33 Pa and a temperature of 300 K after final cleaning. After some of the substrates were ruled out, the reproducibility of forming a monomolecular layer of organic contaminants on an initial substrate surface amounted to 99.5%.

3.5 ANALYSIS OF PLASMA PARTICLES IMPINGING ON THE SURFACE BEING TREATED

This monograph discusses treating surface materials with low-temperature plasma generated by a high-voltage gas discharge outside the electrode gap in a nonuniform electric field in which negatively charged particle fluxes (negative ions and electrons)

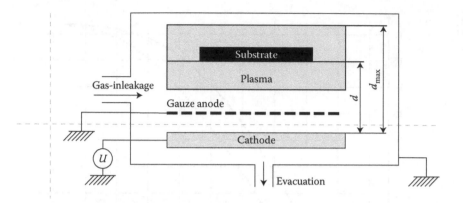

FIGURE 3.11   Schematic of the reactor.

are generated that are incident on the surface being treated. Figure 3.11 shows the schematic of the reactor used for this purpose.

To determine what interaction mechanisms take place between plasma particles and the surface, we will briefly analyze the processes that occur in high-voltage gas-discharge plasma. It is known that the progress of a process occurring in gas-discharge plasma depends on the mean free path and the number of collisions, as well as on the particle energy. The particle energy is, in turn, determined by the gradient of voltages applied to the electrodes of the gas-discharge device, whose action in a nonuniform electric field is a function of distance. This relationship is presented in Reference 65 as

$$y = \frac{hU_n}{U} + 2c\,tg\left(\frac{\pi U_n}{2U}\right),\tag{3.4}$$

where U_n is the electric-field potential at a distance of y from the cathode (V). For purposes of calculating U_n outside the electrodes, we can disregard the first term in Equation 3.4 (because it is small as compared with the second one) and express U_n as follows:

$$U_n = \frac{2U}{\pi}\,arc\,tg\left(\frac{y}{2c}\right).\tag{3.5}$$

Now if we assume that y is the mean free path of the charged particle (λ) multiplied by the number of its collision (n) and substitute this product into Equation 3.5, then U_n will represent the potential difference passed through by the particle after each collision. Then the accelerating potential difference after each collision can be described by

$$\Delta U_n = U - \frac{2U}{\pi}\,arc\,tg\left(\frac{n\lambda}{2c}\right) = U\left(1 - \frac{2}{\pi}\,arc\,tg\left(\frac{n\lambda}{2c}\right)\right).\tag{3.6}$$

Taking into account the energy lost by the particle in collision with process-gas molecules, the particle's full energy after each collision is given by

$$E_n = E_{n-1}(1-\gamma) + e\Delta U_n, \tag{3.7}$$

where $\gamma = 4mM/(m+M)^2$; m is the weight of a charged particle and M is the weight of a process-gas molecule.

Note that n varies from one to the number of collisions that the particle has time to undergo in covering the distance from the cathode to the sample surface ($d = 4.5$ cm in the case under study). With expression (3.5), it is also possible to calculate the maximum distance over which the ion-plasma flux propagates at a given electrode voltage. Assuming that $U_n = U - U_1$, where U_1 is a penalty function (obtained from a natural experiment) and a constant equal to $U_1 = 1$ V, we obtain the following expression:

$$d_{max} = 2c\ tg\left[\frac{\pi}{2}\left(\frac{U-U_1}{U}\right)\right], \tag{3.8}$$

where d_{max} is the maximum distance over which the ion-plasma flux propagates. Expression (3.8) is valid because as the accelerating voltage increases, the distance d_{max} increases as well (this relationship was visually observed during the experiment). Therefore, an essential condition to starting a treatment with high-voltage gas-discharge off-electrode plasma is

$$d_{max} > d. \tag{3.9}$$

Simultaneous analysis of expressions (3.6) and (3.7) shows that in the case of a large number of collisions and a small free path (conditions typically observed at low electrode voltages on the order of the ignition voltage), E_n falls within the energy range typical of plasma-chemical etching (see Figure 3.12). From the graph in Figure 3.12 it follows that as the voltage increases, E_n increases too. At a certain electrode voltage, E_n is sufficient for physical sputtering as well (ion-plasma etching). And if the charged particles are chemically active toward the material being treated, material removal due to chemical reactions occurs along with sputtering—that is, ion-chemical etching takes place.

According to the law of distribution of equipotential lines (see Figure 1.2), the accelerating voltage reduces in the direction of d_{max} at a given fixed voltage at the gas-discharge device's electrodes. This means that by moving the sample from the cathode along the discharge direction, one can change the accelerating voltage in the treatment area. To think of it another way, if a given voltage is applied to the electrodes that exceeds the ignition voltage of high-voltage gas discharge and if the sample is moved along the plasma flux, one can always determine the areas of ion-chemical etching and plasma-chemical etching as well as the transition area (see Figure 3.13), whose etching mechanism will be discussed below.

Let us consider what particles will collide with the surface being treated in each of these areas, assuming by way of illustration that fluorinated gas is used as plasma-forming gas.

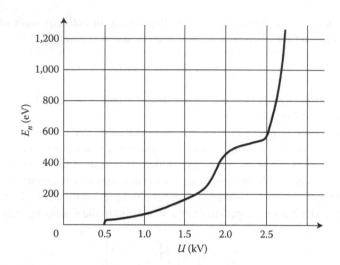

FIGURE 3.12 Relationship between the energy of negative ions bombarding the sample and the electrode voltage at a discharge current of $I = 140$ mA.

In the plasma-chemical etching area, the following particles contribute to the process:

- Negative F⁻ ions (flux J_i^-) generated in plasma as a result of electron attachment to the neutral fluorine atoms [92] under whose action chemically active particles (CAPs) are desorbed from the surface being treated (the desorption is characterized by the desorption coefficient k_2) and material etching takes place (the etching is characterized by the coefficient k_1).
- Chemically active neutral fluorine radicals (CAPs in the flux J_a), which can form reaction products that are volatile at the process temperature when colliding with the atoms of the material being treated. (Only those negatively charged particles move toward the surface that are not passive toward the material being treated [62].)
- Passivating particles (flux J_p). (Their flux incident on the surface is negligible because the surface being treated is outside the discharge-generating

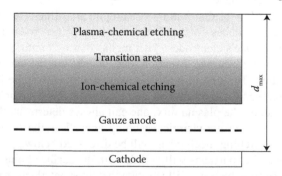

FIGURE 3.13 Arrangement of etching areas in high-voltage gas-discharge plasma.

electrodes, which in itself excludes the generation of passivating particles as a result of cathode sputtering and therefore their effect on etching in general.)

- Neutral molecules of fluorinated gas (flux J_n), whose dissociation under the action of ion bombardment results in chemically active fluorine radicals forming at the surface.

In the ion-chemical etching area, only negative F^- ions (flux J_i^-) will impinge on the surface during etching. In addition to physical sputtering characterized by the sputtering coefficient k_3, these ions will also etch the material because of the chemical reaction taking place. Given the heating of the material by high-energy F^- ions, which reduces the number of collisions between CAPs and the surface because of desorption, we can disregard the effect that CAPs have on etching in this area.

3.6 MECHANISM OF SURFACE CLEANING WITH DIRECTED FLUXES OF LOW-TEMPERATURE OFF-ELECTRODE PLASMA

3.6.1 CLEANING MECHANISM

The substrate surfaces were cleaned with high-voltage gas-discharge plasma. As noted earlier, in this plasma, only negatively charged particles move toward the surface. When the process gas is air containing 78% nitrogen and 21% oxygen, the particles will be represented by negative oxygen ions and electrons. Positive nitrogen and oxygen ions will move toward the gas-discharge device's cathode.

At high gas pressures in the working chamber (i.e., at low accelerating voltages in the range $0.5 \le U \le 1$ kV), the substrate surface will invariably have an adsorbed layer of neutral molecules of NO, N_2O, and NO_2, which result from the recombination of oxygen ions and nitrogen ions [66], and of N_2 and O_2 molecules. Without additional activation, chemical reactions of these compounds with the molecules of organic contaminants (hydrocarbons $-C_xH_y-C_mH_n-$) either do not take place or are weak.

With ion bombardment, contaminants are removable with the technique for plasma-chemical etching described in detail in References 209 and 210 and in Chapter 5. Under the action of ion bombardment, the neutral molecules of NO, N_2O, NO_2, N_2, and O_2 dissociate at the surface to form positive nitrogen and oxygen ions, which begin to move along the electric-field lines toward the cathode, and oxygen radicals, which actively interact with free surface hydrocarbon bonds. The oxidation of organic contaminants occurs in a chain of reactions, apparently in the following steps [190]:

$$RH + O \rightarrow \dot{R} + \dot{O}H; \quad R\dot{O}O + RH \rightarrow ROO\dot{H} + \dot{R};$$

$$RH + O \rightarrow \dot{R}' + \dot{R}''O; \quad ROOH \rightarrow \dot{R}O + \dot{O}H;$$

$$RH + \dot{O}H \rightarrow \dot{R} + H_2O; \quad \dot{R}_nO \rightarrow R_{n-1}O' + RO;$$

$$\dot{R} + O_2 \rightarrow ROO'; \quad 2\dot{R} \rightarrow R_2,$$

where R and Ṙ are the fragments and radicals of organic-contaminant molecules and R′ and R″ are radicals derived from R and formed during the breakdown of contaminant molecules. As a result of such reactions, the film of organic contaminants is broken down into individual fragments that have a low-molecular mass, and then the fragments are oxidized, yielding CO and CO_2, as well as H_2O vapor.

At an accelerating voltage of $U > 1$ kV, the removal of organic contaminants will occur under the mechanism of ion-chemical etching described in detail in References 209 and 210, and in Chapter 5. But as can be seen from Figure 3.14, at an irradiation time of up to 50 s, a discharge current of up to 10 mA, and an accelerating voltage of up to 3 kV, the substrate temperature does not exceed 360 K [211]. Therefore, adsorption of neutral molecules at the surface is possible, and the ion-chemical etching mechanism of contaminants differs from the mechanism discussed earlier in that the neutral molecules of NO, N_2O, NO_2, N_2, and O_2 are present at the surface that serve as sources for the formation of chemically active oxygen radicals. From this it follows that contaminants will be removed not only by the ion-chemical etching described in References 209 and 210 but also by the mechanisms of ion-stimulated etching and electron-stimulated radical etching [212,213].

Contaminant removal comprises the following key processes:

1. Physical sputtering with negative oxygen ions
2. Chemical etching with negative oxygen ions
3. Chemical etching with oxygen radicals formed through the dissociation of neutral molecules as a result of bombardment with negative oxygen ions

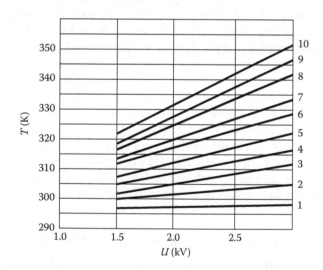

FIGURE 3.14 Experimental relationship between substrate temperature and accelerating voltage: (1) $I = 1$ mA, (2) $I = 2$ mA, (3) $I = 3$ mA, (4) $I = 4$ mA, (5) $I = 5$ mA, (6) $I = 6$ mA, (7) $I = 7$ mA, (8) $I = 8$ mA, (9) $I = 9$ mA, and (10) $I = 10$ mA. Irradiation time $t = 50$ s.

4. Chemical etching with oxygen radicals formed through the dissociation of neutral molecules as a result of electron impact

Physical sputtering of organic-contaminant particles causes unsaturated valence bonds to form that are highly chemically active and thus intensify the chemical reactions to a greater degree.

From References 190 and 214, it is known that hydrocarbons are capable of polymerization when exposed to a plasma flux. The polymerization rate is proportional to the process temperature [215]. As a result of polymerization, the molecules of organic contaminants merge into long, cross-linked chains. The strengthening of molecular bonds impedes the removal of organic-contaminant particles from the substrate surface and slows down the cleaning process. According to Reference 215, the most efficient polymerization occurs at a temperature equal to or greater than 400 K. But at a process temperature of 360 K or below, the polymerization rate does not exceed 0.96% per hour; therefore, for purposes of this monograph, polymerization is negligible. In this case, the total contaminant-removal rate depends on the rates of the processes discussed above.

3.6.2 CLEANING MODEL: PRIMARY EXPRESSIONS

We will determine surface cleanliness as the relationship between the variation value of the surface concentration of contaminants, the contaminant-removal rate (V_{et}), and the irradiation time (t):

$$C_d = C_{(0)d} - V_{et}\rho t, \tag{3.10}$$

where $C_{(0)d} = h \cdot \rho$ is the initial concentration of surface contaminants; h is the thickness of the contaminant film; and ρ is the contaminant density. We propose that V_{et} be expressed as a sum of rates:

$$V_{et} = V_{ich} + V_{ist} + V_{est}, \tag{3.11}$$

where V_{ich} is the ion-chemical etch rate; V_{ist} is the ion-stimulated radical etch rate; and V_{est} is the electron-stimulated radical etch rate. Considering the presence of neutral molecules at the surface, the ion-chemical etch rate is given by [209]

$$V_{ich} = \left(\frac{B(k_1 + k_3)M}{\rho N_A}\right)\left|\exp\left(\frac{U - U_{gd}}{U}\right) - 1\right| \times J_i^-(1 - \theta), \tag{3.12}$$

where B is the value of the penalty function obtained from a natural experiment (a constant); k_1 and k_3 are the coefficients of plasma-chemical etching and physical sputtering, respectively; M is the molar mass of organic contaminants; U_{gd} is the electrode voltage at which the ion energy lies at the boundary between the energies of plasma-chemical etching and ion-chemical etching at the moment the ion approaches the surface being treated; J_i^- is the flux of negative ions incident on the substrate; and θ is the degree of surface filling by active particles. The value of J_i^- is given by [209]

$$J_i^- = \left(1 - \frac{d}{d_{max}}\right)\frac{I}{qeS_K}\left(1 - \frac{\gamma_e\,\eta}{(1+\gamma_e)}\exp[(\alpha - \alpha_n)d_{max}]\right), \tag{3.13}$$

where $(1 - d/d_{max})$ is a multiplier that numerically characterizes those ions from the aggregate flux that reach the sample surface and that contribute to the etching, as long as $d_{max} > d$, where d_{max} is the maximum distance over which the ion-plasma flux propagates, and d is the distance to the substrate; S_K is the cathode surface area; q is the geometric transparency of the gauze anode; and γ_e, α, and α_n are the coefficients of secondary emission, ionization, and attachment, respectively. Dissociation of four molecule types—NO, N$_2$O, NO$_2$, and O$_2$—generates chemically active oxygen radicals at the surface. Therefore, the total degree of surface filling by active particles equals

$$\theta = \theta_{NO} + \theta_{N_2O} + \theta_{NO_2} + \theta_{O_2}. \tag{3.14}$$

If plasma does not contain passivating particles, $\theta_{NO}, \theta_{N_2O}, \theta_{NO_2},$ and θ_{O_2} can be expressed as

$$\theta_{NO} = 1/\left[1 + \frac{(k_1 + k_2^{NO})\,J_i^-}{s_a\,J_n^{NO}}\right], \quad \theta_{NO_2} = 1/\left[1 + \frac{(k_1 + k_2^{NO_2})\,J_i^-}{s_a\,J_n^{NO_2}}\right],$$

$$\theta_{N_2O} = 1/\left[1 + \frac{(k_1 + k_2^{N_2O})\,J_i^-}{s_a\,J_n^{N_2O}}\right], \quad \theta_{O_2} = 1/\left[1 + \frac{(k_1 + k_2^{O_2})\,J_i^-}{s_a\,J_n^{O_2}}\right], \tag{3.15}$$

where $k_2^{NO}, k_2^{N_2O}, k_2^{NO_2}$, and k_2^{O} are desorption coefficients for NO, N$_2$O, NO$_2$, and O$_2$ molecules calculated with the method presented in Reference 214:

$$k_2^i = \frac{3}{4\pi^2}\beta_i\frac{4mM_i}{(m+M_i)^2}\frac{E_n}{E_{sv}}, \tag{3.16}$$

where β_i is a unitless parameter dependent on the mass ratio of the incident ion (m) and the target particle of the i type of molecule (M_i); E_n is the ion energy at the moment the ion approaches the surface; and $J_n^{NO}, J_n^{N_2O}, J_n^{NO_2}$, and $J_n^{O_2}$ are neutral-molecule fluxes incident on the substrate surface, and they are calculated with the methods presented in References 190 and 216. The aggregate coefficient of physical sputtering is given by

$$k_3 = k_3^C + k_3^H, \tag{3.17}$$

where k_3^C and k_3^H are the coefficients of physical sputtering derived from Equation 3.16 for carbon and hydrogen atoms, respectively.

Given these calculations, the ion-stimulated etch rate can be determined from [212]

$$V_{ist} = \frac{Bk_i^* k_1 M}{\rho N_A} J_i^- \theta, \tag{3.18}$$

where k_i^* is the unitless coefficient of ion-stimulated etching. Similar to Equation 3.18, the electron-stimulated etch rate can be expressed as [213]

$$V_{est} = \frac{Bk_e^* k_1 M}{\rho N_A} J_e \theta, \tag{3.19}$$

where k_e^* is the unitless coefficient of electron-stimulated etching and J_e is the electron flux incident on the substrate surface, and it is derived from the following expression [209]:

$$J_e = \frac{I \gamma_e \eta}{eq S_K (1 + \gamma_e)} \exp[(\alpha - \alpha_n) d_{\max}]. \tag{3.20}$$

By substituting Equation 3.17 into Equation 3.12 and Equations 3.12, 3.18, and 3.19 into Equation 3.11 and transforming the equations, we obtain the final expression for determining the aggregate removal rate for organic contaminants:

$$V_{et} = \frac{BM}{\rho N_A}$$
$$\times \left[(k_1 + k_3^C + k_3^H) \left| \exp\left(\frac{U - U_{gd}}{U} \right) - 1 \right| J_i^- (1 - \theta) + k_i^* k_1 J_i^- \theta + k_e^* k_1 J_e \theta \right]. \tag{3.21}$$

Given Equations 3.11 and 3.21, the expression for determining the variation value of the surface concentration of organic contaminants can now be expressed as [217,218]

$$\bar{C}_d = \rho h - \frac{BM}{N_A}$$
$$\times \left[(k_1 + k_3^C + k_3^H) \left| \exp\left(\frac{U - U_{gd}}{U} \right) - 1 \right| J_i^- (1 - \theta) + k_i^* k_1 J_i^- \theta + k_e^* k_1 J_e \theta \right] t. \tag{3.22}$$

The proposed analytical relationship makes it possible to reduce the volume of expensive and complex research into the effect that the parameters of ion–electron bombardment have on cleaning. The calculation results obtained with expression (3.22) can be directly used in technology for fabricating DOEs and nano- and micro-electronic elements.

3.7 EXPERIMENTAL INVESTIGATION INTO THE RELATIONSHIP BETWEEN THE DEGREE OF SURFACE CLEANLINESS AND PHYSICAL PLASMA PARAMETERS

Experimental variation values of the surface concentration of organic contaminants were measured with the tribometric method based on the relationship between the probe's sliding speed on the test surface and the concentration of organic contaminants adsorbed at the surface [70] (for details, see Chapter 2).

Figures 3.15 through 3.17 show experimental graphs that represent the relationship between the surface concentration of organic contaminants on plasma-cleaned substrates and three process factors: accelerating voltage, irradiation time, and discharge current [181,219].

The curves in Figure 3.15 show that, depending on the discharge current, two process types may take place in the treatment area at the surface. At a low discharge current ($0.5 \leq I \leq 1$ mA; see curves 1 and 2), the plasma is observed to be deficient in negative oxygen ions because the concentration of neutral molecules in the process gas is low.

At $0.5 \leq I \leq 1$ mA, the mean free path of charged particles across the entire range of accelerating voltages reaches the value of the distance between the electrodes and the substrate surface [66].

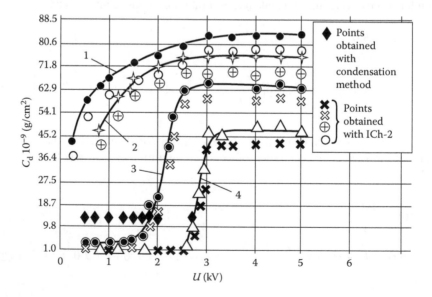

FIGURE 3.15 Relationship between the degree of surface cleanliness and the accelerating voltage: (1) 0.5 mA, (2) 1 mA, (3) 2 mA, and (4) 3 mA. $t = 10$ s.

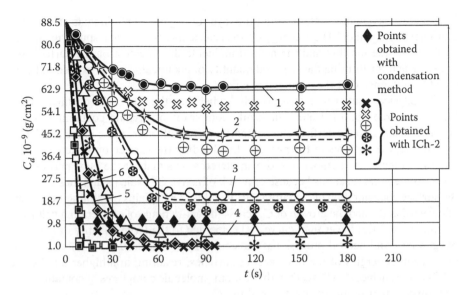

FIGURE 3.16 Relationship between the degree of surface cleanliness and the irradiation time: (1) 0.5 mA (t_{lim} = 65 s), (2) 1 mA (t_{lim} = 60 s), (3) 1.4 mA (t_{lim} = 50 s), (4) 1.5 mA (t_{lim} = 30 s), (5) 2.6 mA (t_{lim} = 20 s), and (6) 3 mA (t_{lim} = 10 s). U = 3 kV. Dashed lines indicate the relationship (3.22).

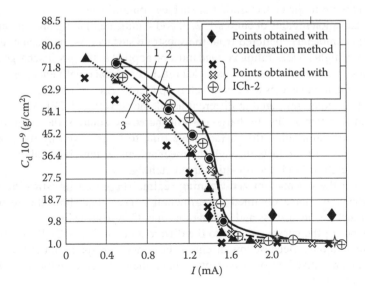

FIGURE 3.17 Relationship between the degree of surface cleanliness and the discharge current at U = 3 kV: (1) t = 50 s, (2) t = 60 s, and (3) t = 120 s.

Negative oxygen ions interact with the layer of organic contaminants at energies that virtually equal eU. High-energy O^- ions cause intense physical sputtering of the surface. But the ion flux and the neutral-molecule flux incident on the surface are not sufficient for forming the required quantity of chemically active oxygen radicals capable of intensifying the chemical reactions.

The physical-sputtering rate is notably higher than the rate of the chemical reactions. Sputtered modified atom–molecule complexes of organic contaminants are adsorbed back onto the surface and contaminate it. With an increase in the accelerating voltage, the energies of the particles bombarding the surface only increase, and there is an increased number of those sputtered atoms that for the most part do not undergo chemical reactions because of a lack of active O^* radicals and thus increase the surface concentration of contaminants. It is this that explains the similar behaviors of curves 1 and 2 and the growth of C_d across the entire range of accelerating voltages.

But curve 2 lies lower than curve 1. This is due to the large quantity of negative oxygen ions and neutral process-gas molecules in plasma (at $I = 1$ mA) and to the diminishing energies of the bombarding particles, resulting in a higher number of CAPs at the surface. CAPs react with the atom–molecule complexes of contaminants and impede their sputtering by diminishing C_d.

Analysis of curves 3 and 4 confirms these statements. At a discharge current of $I \geq 2$ mA, the number of CAPs at the surface is sufficient for the ion-chemical, ion-stimulated, and electron-stimulated etching of organic contaminants. The rate of the chemical reactions is comparable with or exceeds the physical-sputtering rate of organic contaminants. It is this that explains the minimum values of C_d in the voltage ranges $0.5 \leq U \leq 1.2$ kV (curve 3) and $0.5 \leq U \leq 2.5$ kV (curve 4). At voltages of $1.2 \leq U \leq 2.5$ kV (curve 3) and $2.5 \leq U \leq 3$ kV (curve 4), the particle energy reaches a value at which the physical-sputtering processes begin to intensify, resulting in a drastic increase in the concentration of surface contaminants.

The saturation regions of curves 1, 2, 3, and 4 indicate that organic contaminants have been completely sputtered from the substrate surface. As the discharge current increases, a higher accelerating voltage must be applied to the gas-discharge device's electrodes to ensure the charged particles reach the surface with energies sufficient for contaminant molecules to be completely sputtered. As a result, curves 1, 2, 3, and 4 begin to become saturated at different accelerating voltages. Different saturation levels indicate that as the discharge current increases, some of the atom–molecule complexes contaminating the surface have time to be removed by the evacuation system as gaseous products through the mechanisms of ion-chemical etching, ion-stimulated etching, and electron-stimulated etching.

To select the optimal surface-cleaning regime (based on investigating its time dependence), we have obtained experimental curves that show the relationship between surface cleanliness and irradiation time, $C_d = f(t)$ (see Figure 3.16). The different slopes of the curves are due to the different quantities of CAPs reacting with contaminant atoms. Because low discharge currents correspond to the low fluxes of radicals and negative ions of oxygen incident on the surface, removing contaminants takes a longer time.

If there is a CAP deficiency, oxygen radicals will first react with chemically active hydrogen atoms. Removing chemically inert carbon atoms to a level of

$C_d = 10^{-9}$ g cm^2 is possible if the CAP deficiency at the substrate surface is eliminated. This, in turn, is possible at a discharge current of $I = 2.6$ mA and $I = 3$ mA. In the latter case, the irradiation time at which $C_d = 10^{-9}$ g/cm^2 is achieved is as little as 10 s. A long irradiation time at an accelerating voltage greater than 1 kV causes the substrate surface to heat to temperatures that impede the adsorption of neutral process-gas molecules and, therefore, the formation of CAPs [209]. Each value of the discharge current has its own critical irradiation time ($t_{\lim}$), at which the surface heats up to a temperature at which neutral molecules are desorbed from it. The value of θ tends to zero in the region $t \geq t_{\lim}$, thereby reducing the removal rate of organic contaminants at low currents. This explains the segments of curves 1, 2, 3, and 4 (Figure 3.16) in which the value of C_d ceases to depend on plasma-treatment time. These segments can likely be explained as well by the presence of a hard-to-remove modified layer of carbon compounds whose removal requires high discharge currents. The value of $t_{\lim}$ for each discharge current is determined by the intersection point of tangents to the appropriate curve (see Figure 3.16).

The relationship $C_d = f(I)$ in Figure 3.17 shows that for three different cleaning times, a sharp reduction in the contaminant concentration at the surface is observed for all curves in the range $1.4 \leq I \leq 1.6$ mA. The sharp decrease in the value of C_d in this discharge-current range is due to the intense filling of the surface by chemically active O* radicals formed as a result of the ion and electron stimulation of neutral process-gas molecules as well as ion-chemical etching [209]. The CAP deficiency at the surface typical of lower discharge currents ($I \leq 1.4$ mA; see also Figure 3.15) begins to be eliminated. At $I \geq 2.6$ mA, C_d reaches its minimum value, and this confirms that the CAP deficiency is fully eliminated.

AFM images showing the surface before (Figure 3.10) and after (Figure 3.18) final cleaning confirm that high-voltage gas-discharge off-electrode plasma efficiently removes organic contaminants from substrate surfaces.

If the initial surface has a contaminant film and shows hydrophobic properties (the interaction distance between the cantilever and the surface structure is a 3-nm

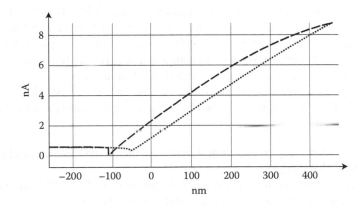

FIGURE 3.18 Force curve obtained with Solver PRO-M for the test surface after final cleaning. The interaction distance between the cantilever and the surface structure is an H$_2$O film, $H = 20$ nm.

film of organic contaminants; see Figure 3.10), after final cleaning, the surface shows hydrophilic properties and therefore actively adsorbs water. Because of capillary action [203,206], the layer thickness H equals 20 nm. Therefore, the curve in Figure 3.18 characterizes a technologically clean surface because the surface is free from molecules of organic contaminants.

Thus, to attain the precision-cleanliness level, this monograph recommends cleaning substrate surfaces with high-voltage gas-discharge plasma at $U = 1.2$ kV, $I = 3$ mA, and $t = 10$ s.

The experimental contaminant concentrations were obtained with the tribometric method, which is discussed in Chapter 2. The method's reliability was confirmed by using the condensation method to take similar measurements with the ICh-2 tribometer (see Figures 3.15 through 3.17). The variance of the readings obtained from the methods under the same regimes was found to not exceed 10%. The higher concentrations obtained with the condensation method are due to its threshold sensitivity equal to 10^{-8} g/cm^2, while the low C_d obtained with the ICh-2 tribometer is due to error caused by the probe breaking away from the surface, poor resistance reproducibility of the signal contacts, and friction in moving parts [109].

For comparison, Figure 3.16 shows the curves derived from expression (3.22). The close agreement between the curves and the experimental data shows that the proposed model adequately describes the process of physically cleaning organic contaminants off substrate surfaces with high-voltage gas-discharge off-electrode plasma. An important achievement is that using the proposed cleaning method in technology for fabricating subminiature relays for space applications and semiconductors helped improve the contact conductivity of the former [220] and the parameters of the latter, which were measured with the quick-measurement unit described in detail in Reference 221.

3.8   PROCEDURE FOR FINAL SURFACE CLEANING WITH OFF-ELECTRODE PLASMA

The procedure for final cleaning of substrate surfaces with off-electrode plasma consists of several consecutive steps that, once completed, yield a technologically clean surface (a surface with a cleanliness level of 10^{-9} g/cm^2). Coupled with vacuum techniques for mask deposition that are completed as part of a single cycle, this procedure helps fabricate microstructures with desired parameters.

When a mask is deposited through the thermal vacuum evaporation technique, the final-cleaning procedure is completed as part of that technique and comprises the following steps:

1. Placing chemically cleaned initial substrates on the substrate carousel in the vacuum chamber
2. Evacuating the vacuum chamber to a pressure of $7.5 \cdot 10^{-2}$ Torr with a mechanical pump; configuring the high-voltage gas-discharge device: setting the discharge current at 3 mA, the electrode voltage at 1.2 kV, and the irradiation time at 10 s (per substrate)

3. Cleaning the initial substrates under these conditions with a low-temperature plasma flux to a level of 10^{-9} g/cm^2
4. Deactivating the high-voltage gas discharge
5. Depressurizing the working chamber
6. Removing the substrates from the carousel and placing them in the desiccator or the tribometer

This final-cleaning procedure makes it possible to obtain a technologically clean surface.

3.9   CHAPTER SUMMARY

The existing methods for cleaning substrate surfaces have several major drawbacks:

- The methods cause surface defects in the form of craters and wavelike relief.
- HF and SHF discharge parameters that are difficult to stabilize and the loading effect undercut efficiency.
- The equipment used is complex, energy-intensive, and expensive.

Our investigations based on the proposed gas-discharge devices enable us to propose efficient methods for fabricating surfaces with the desired properties, including a surface-cleaning method. We have experimentally proved that the cleaning method yields a surface with a cleanliness level of 10^{-9} g/cm^2 and stands out from the others thanks to its low cost and energy consumption.

The results we obtained can be used in technology for fabricating DOEs and nano- and microelectronic elements.

4 Adhesion in Metal–Dielectric Structures after Surface Bombardment with an Ion–Electron Flux

Thin metal films (up to 50–100 nm thick) are widely used as masks in vacuum-plasma etching of nano- and microelectronic components—for example, in fabricating trenches [34] and DOE microreliefs [1,2]. Stringent requirements for trenches [34] and DOE microreliefs [1,2] pose the problem of enhancing mask adhesion: even an insignificant decrease in adhesion leads to a significant undercut of the substrate material along the metal–substrate interface, thereby irreversibly distorting the shape of the structure.

The problem of adhesion (A) has been dealt with in quite a number of publications [222–226], and this indicates that the problem remains highly relevant. But comprehensive studies of adhesion mechanisms are lacking that would treat problems such as the effect that the parameters of gas-discharge plasma have on adhesion and the effect of adhesion on the geometric parameters of micro- and nanostructures. Several publications [227,228] attribute the gap to the fact that the adhesion strength between atoms and metal and dielectric molecules depends on the strength of their chemical bonds, Van der Waals forces, the presence of intermediate layers, chemical interaction, surface cleanliness, and so forth. These conditions make it extremely difficult or infeasible to accurately quantify adhesion strength and explain adhesion increase and decrease mechanisms.

Mask adhesion mostly depends on the cleanliness of substrate surfaces. References 229 and 230 indicate that an increase in the strength of metal film–substrate adhesion relates not only to surface cleaning but also to surface modification. Reference 231 investigates the effect that the treatment of a SiO_2 surface with glow-discharge plasma has on the adhesion of metal condensates to SiO_2. The adhesion mechanism of metal films predeposited on a dielectric surface and treated with an ion–electron flux has not been studied before.

4.1 ADHESION-ENHANCING MECHANISM

The studies were conducted on copper–polycrystalline glass and chromium–silicondioxide structures. Metal films were deposited on the substrates through the thermal vacuum evaporation technique [232] at a pressure of $5 \cdot 10^{-6}$ Torr. The films' thickness was checked with the MII-4 and NewView 5000 microinterferometers and varied between 50 and 100 nm. Before the metal films were deposited, the surfaces were chemically pretreated with the technique described in Section 3.4, including the final predeposition cleaning with an ion–electron flux at an accelerating voltage of 1.2 kV, a discharge current of 3 mA, and an exposure time of 10 s, in the air atmosphere.

The structures under analysis were bombarded with an ion–electron flux generated by a high-voltage gas discharge outside the electrode gap. Identical initial properties of the substrate surfaces were obtained through the controlled-contamination technique [71]. The metal films' adhesion was measured with the normal-detachment technique [233]. To ensure the reliability and reproducibility of the experimental results, each treatment mode was reproduced on 10 identical samples. The spread between the average maximum and minimum values of the adhesion strength (ΔA) was no greater than 15%.

It is known [109,234,235] that, when placed into the working chamber, even technologically clean substrates become contaminated by hydrocarbons such as photoresist molecules and organic acid salts that are formed at the forevacuum stage when the thermal vacuum evaporation technique is used to form masks. A distinctive feature of hydrocarbon compounds is their common base represented by the C_xH_y structure [236]. So, to simulate an organic contaminant, we used monomolecular layers of vacuum oil, which is presently best studied and allows the investigation results to be reliably interpreted. For purposes of this monograph, we studied the metal–organic contamination–substrate (Me–C_xH_y–Sub) and metal–substrate (Me–Sub) structures (see Figure 4.1), whose surfaces were bombarded with an ion–electron flux.

In the Me–C_xH_y–Sub structure before bombardment, the adhesion magnitude is governed by physical absorption: the adhesion strength between the free bonds of atoms and the molecules of organic contaminants, and the analogous bonds of metal film and substrate atoms. The surfaces in contact have a low value of free surface energy [117].

When the Me–C_xH_y–Sub structure is treated with an ion–electron flux until a meltforms [91], gas bubbles appear on its surface at the initial instant of time. The bubbles are detectable with MII-4 as microswellings in the thin film after a sharp cooling ($\approx$20°C per minute) (see Figure 4.2a).

The microswellings indicate that at the Me–C_xH_y–Sub interface the bombardment finally results in irreversible processes of full organic-molecule dissociation. As a result, the interface contaminants turn into a gaseous product and a precipitate

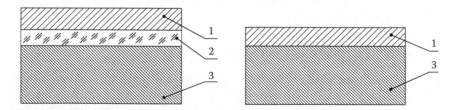

FIGURE 4.1 Structure of the test samples: (1) metal, (2) organic-contamination layer, and (3) substrate.

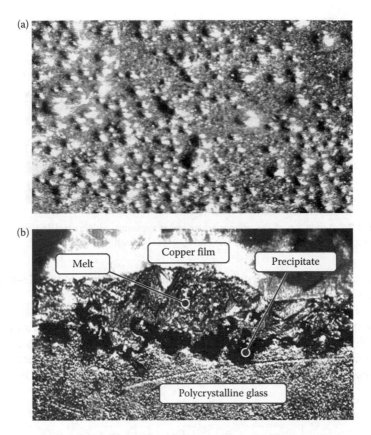

FIGURE 4.2   A copper (Cu)–chemical compound–polycrystalline glass (ST-50-2) structure was briefly ($t_{ir} \approx$ 1.5–2 s) brought up to the metal film's melting point and then sharply cooled ($\approx$20°C per minute): (a) appearance of the thin-film surface, ×36, and (b) angle lap of the copper–chemical compound–polycrystalline glass structure in the microswelling region of the film, ×400.

(see Figure 4.2b). The precipitate was also detected for the treatment mode of $T < T_m$ (where T_m is the melting point). Note that as distinct from the previous case, the precipitate was distributed uniformly (see Figure 4.3): because the metal film was not deformed, the energy was uniformly distributed over the entire interface. Therefore, we conclude that bombarding the surface of the metal–organic contamination–substrate structure with an ion–electron flux is conducive to an active dissociation of hydrocarbon molecules. The chemically active radicals of the gas and the precipitate interact with the surface atoms of the substrate and the metal film, thereby forming oxides, nitrides, and carbides [237] characterized by a significant value of the binding energy [238].

Therefore, the formation of the metal–chemical compound–substrate system (Me–$C_x^*H_y^*$–Sub) is a major mechanism for enhancing adhesion in the structure under study (see Figure 4.1). To achieve a maximum adhesion strength, we need to find out in which way it depends on the surface-contamination parameter and to

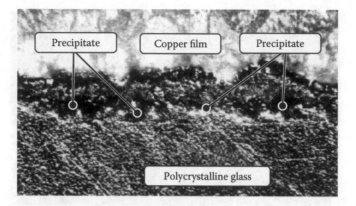

FIGURE 4.3 Angle lap of the copper–chemical compound–polycrystalline glass structure after treatment at $I = 100$ mA, $U = 2$ kV, and $t_{ir} = 3$ min, ×400.

carry out a comparative analysis of the relevant characteristics of the Me–C_xH_y–Sub structures before and after ion–electron bombardment.

For convenience, the time that substrates are exposed to contamination, t_{cont}, is taken as the contamination parameter. Comparing the curves of the function $A = f$ (t_{cont}) in Figure 4.4 suggests that the maximum values of the metal film's adhesion to the substrate surface for the untreated (curve 1) and treated (curve 2) metal–organic contamination–substrate structures are observed at $t_{cont} = 3$ min. This may be

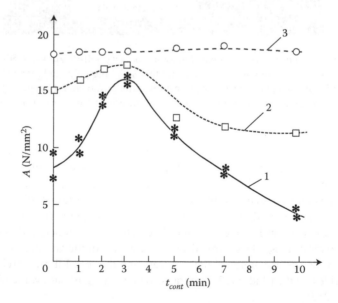

FIGURE 4.4 Strength of the chromefilm's adhesion to the surface of the SiO_2 substrate, as a function of the substrate's exposure to contamination: (1) untreated structure; (2) structure treated at $I = 100$ mA, $U = 2$ kV, and $t_{ir} = 5$ min; and (3) substrate adhesion after final cleaning at $I = 3$ mA, $U = 1.2$ kV, and $t = 10$ s.

because the contamination-exposure time allows a monomolecular contamination layer to coat the entire substrate surface. The layer's surface atoms form either chemical bonds with metal and substrate atoms (after bombardment-induced dissociation; curve 2) or Van der Waals force-induced bonds (curve 1) [99]. As a result, interaction between the surface atoms of the hydrocarbons, metal, and dielectric with free bonds occurs over the entire metal–organic contamination–substrate interface. The number of such interactions is restricted by the number of free surface bonds of the organic molecules (N_0). The increase in the adhesion strength at $0 < t_{cont} < 3$ min testifies to the absence of the monomolecular layer during that period and depends on the hydrocarbon molecule-aided growth of adhesion centers in local regions of the substrate surface (see Figure 4.5).

As a result of multiple interactions with the adsorbent, adsorbate particles are located on the adsorption centers such that their binding energy is maximal. Because of this, once the first monomolecular layer is formed, less adsorption-active sides of adsorbate molecules become oriented outside. As a result, the binding force between the first and second layers and between all other adjacent layers becomes significantly weaker [235]. This is the reason why the adhesion value decreases for the contamination-exposure times of $3 < t_{cont} \leq 10$ min—in this case organic-contamination molecules have time to form multilayer conglomerations at the substrate surface. In the subsequent research, however, the contamination time was deliberately taken to be equal to $t_{cont} = 10$ min in order to demonstrate the advantages of using for enhancing mask adhesion an ion–electron flux generated by a high-voltage gas discharge. Greater values of the adhesion strength in the case of bombardment (curve 2)

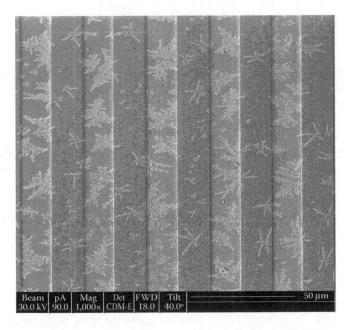

FIGURE 4.5 Hydrocarbon molecule-aided growth of adhesion centers in local regions of the substrate surface. Image obtained with a Philips XL 40 scanning electron microscope.

are characterized by chemical adsorption, which has a greater binding energy in comparison with physical adsorption [115]. The manner in which curve 3 changes confirms that the cleaning method discussed in Chapter 3 produces a technologically clean surface.

4.2   ADHESION MODEL: PRIMARY EXPRESSIONS

The adhesion of a metal film to a substrate surface treated with an ion–electron flux is a function of the irradiation time t_{ir}, the discharge current I, and the accelerating voltage U. To obtain numerical values of adhesion strength, this monograph proposes a model whose primary expression links adhesion to these parameters.

According to Reference 239, the value of A depends on the activation energy (E_a), the number of free bonds on each contacting surface (N), and the energy supplied to the system. These are linked by the expression

$$N = N_0 \exp\left[-\nu\, t \exp\left(-\frac{E_a}{kT}\right)\right],$$ (4.1)

where ν is the valence-vibration frequency of atoms and t is the energy-dissipation time. When the surface is modified with gas-discharge plasma and a metal film is deposited on it, expression (4.1) changes to [231]

$$N = N_0 \exp\left[-\exp\left(-\frac{E_s}{E_{MeO}}\right)\right],$$ (4.2)

where N is the number of broken bonds at the surface of the intermediate layer between the metal film and the oxide substrate and N_0 is the number of bonds at the oxide surface. In Equation 4.2, the input energy is taken to be equal to the energy of the individual bond between the deposited metal and the oxide (E_{MeO}), and the energy of the individual bond in the surface layer of the oxide (E_s) is substituted for the activation energy. From this, it follows that the adhesion strength depends on the number of broken bonds at the surface of the intermediate layer. In the case under study, the organic-contamination layer acts as an intermediate layer. Clearly, it makes sense to use Equation 4.2 to calculate N only at the initial stage—that is, before treating the Me–C_xH_y–Sub system with an ion–electron flux. But during the treatment, the total number of free bonds of organic molecules will equal the following sum:

$$N_{sum} = N + N_{ir},$$ (4.3)

where N_{ir} is the number of free bonds of the organic molecules formed as a result of the treatment. To determine N_{sum}, we must know the value of N_{ir}, which depends on the system's initial state and I, U, and t_{ir}. In this case, the following expression is valid:

$$N_{ir} = N\, f(I,U,t_{ir}).\tag{4.4}$$

Thus, we need to find an expression for the function $f\,(I,\,U,\,t_{ir})$. Let us consider the factors that have an effect on the formation of broken bonds in the hydrocarbon layer during the ion–electron treatment of the metal–organic contamination–substrate structure.

In analyzing Figure 4.1 and drawing an analogy with Equations 4.1 and 4.2, we can assume that $E_s = E_{Me-C_xH_y}$ and $E_{MeO} = E_n(U)$, where $E_{Me-C_xH_y}$ is the energy of the individual bond between the deposited metal and organic contaminants, and $E_n(U)$ is the energy transferred by the particles of the ion–electron flux to the metal film's surface, and the energy is a function of the accelerating voltage [209]. Then

$$f(I,U,t_{ir}) \sim \exp\!\left(-\frac{E_{Me-C_xH_y}}{E_n(U)}\right).\tag{4.5}$$

For fabricating submicron and sub-100 nm structures in microelectronics, lower values of $E_n(U)$ are preferable [34]. In the studies conducted for purposes of this monograph, $E_n(U)$ was no greater than 800–900 eV. At such particle energies, the depth of particles' penetration into the metal is about 10 nm [240]—that is, the energy is transferred to the Me–C_xH_y–Sub structure at the near-surface area of the metal film. Negative ions and electrons do not have a direct effect on the formation of free surface bonds of organic molecules, and models based on the thin-film penetration method [240,241] as well as models taking account of the stress exerted by the ion–electron flux on the surfaces of solids [242] are inapplicable.

Because of this, the proposed model uses a different approach, one that is based on the mechanism of temperature wedges [243]. Under that mechanism, in an environment with microscopic characteristics (heat capacity C, density ρ, and thermal diffusivity D_T), energy E is emitted at the beginning of the system's coordinates at the instance of time $t = 0$ and spreads all across the crystal according to the laws of heat conduction. For the heat-conduction equation under initial and terminal conditions

$$T|_{r=0} = E; \quad \left.\frac{\partial T}{\partial r}\right|_{r=L} = -\frac{E}{D_T}; \quad T|_{t=0} = \begin{cases} E, & \text{for } r = 0, \\ T_0, & \text{for } 0 < r \le L \end{cases}$$

a solution is given by [244,245]

$$T(r,t) = T_0 + \frac{E}{c\rho(4\pi D_T t)^{3/2}}\exp\!\left(-\frac{r^2}{4 D_T t}\right).\tag{4.6}$$

Expression (4.6) presents the conclusion that t is the time required for the temperature balance t_{tb} to be established in the system (the crystal) and that in an environment with microscopic characteristics the exponential factor is negligible.

For the purposes of the approach under discussion, this conclusion helps us look into the Me–C_xH_y–Sub structure. The substrate holder also has an effect on thermal processes. Thus, from the perspective of heat-conduction laws, a new system, metal–organic contamination–substrate–substrate holder, can be considered. The establishment of a temperature balance in this system depends on each of its components because they all have different values of D_T. It is only after the temperature balance has been established and at energies of $E_n(U) > E_{Me-C_xH_y}$ that the complete irreversible dissociation of organic molecules begins to take place. As can be concluded from Equations 4.4 and 4.5, the process is exponential. Otherwise the greater part of the energy (heat) supplied to the metal film will be dissipated from it by the substrate and the substrate holder, while the smaller part will be spent to form free bonds of atoms and molecules of the materials that form the phase interface (Me, C_xH_y, Sub). Because the entire phase interface lacks energy, only partial dissociation takes place here.

Macroelements—substrate holder and substrate—determine the time required for a temperature balance to be established. The time for the substrate holder, however, is negligible because metals display a high thermal diffusivity. Thus, the value of t_{tb} is given by

$$t_{tb} = t_{tb}^{(1)}, \tag{4.7}$$

where $t_{tb}^{(1)}$ is the time required for a temperature balance to be established in the substrate. Detailed analysis of Equation 4.6 shows that the multiplier $\rho(4\pi D_T t)^{3/2}$ in the denominator of the second summand is the body mass; therefore, the product $(4\pi D_T t)^{3/2}$ is the volume of the element through which the heat transfers. Thus, knowing the substrate's geometric dimensions, we can express $t_{tb}^{(1)}$ as

$$t_{tb}^{(1)} = \frac{V_1^{2/3}}{4\pi D_T^{(1)}},$$

where V_1 is the substrate's volume. Then

$$t_{tb} = \frac{V_1^{2/3}}{4\pi D_T^{(1)}}. \tag{4.8}$$

It is known [237,246] that thermal diffusivity is a function of energy:

$$D_T = D_0 \exp\left(-\frac{C}{E}\right), \tag{4.9}$$

where C characterizes the material's energetic properties. But in fact not all energy transferred to the body is absorbed by it: some of it will be dissipated to the environment. With this in mind, Equation 4.9 can be rewritten as

$$D_T = D_0 \exp\left(-\frac{E^*}{E}\right), \tag{4.10}$$

where E^* is the capability of the material to lose energy, and it is determined by the material's properties and the interaction between the charged particles of the ion–electron flux and the surface being treated at a given value of energy (E) supplied to it.

Because $D_T^{(1)} \ll D_T^{(2)}$, the greater part of energy loss will depend on the substrate's size and material properties. The numerical value of E^* is determined by the substrate's temperature immediately at the time that a temperature balance is established, and the value is calculated with Equation 4.10 substituted into Equation 4.6:

$$E^* = -E \ln\left[\frac{1}{4\pi D_0^{(1)} t_{tb}}\left(\frac{E}{c\rho(T - T_0)}\right)^{2/3}\right]. \tag{4.11}$$

The relationship $E^* = f(E)$ is shown in Figure 4.6.

This relationship presents the conclusion that with an increase in the amount of supplied energy, the loss dramatically diminishes, and vice versa. At high E^*, the thermal diffusivity is minimal and approximately equal to D_0, while at low E^*, the thermal diffusivity tends to its maximum value, as can be seen from Equation 4.10.

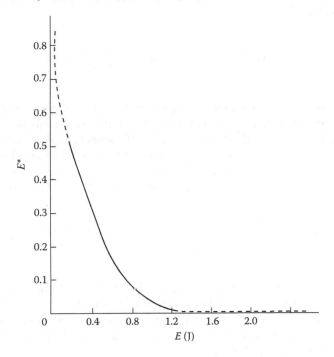

FIGURE 4.6 Relationship between E^* and the energy supplied to the metal–organic contamination–substrate system. Dashed lines indicate the extrapolated values of E^*.

This observation agrees with established opinion [237,246]. Taking into account the above statements, we will write the expression for the thermal diffusivity of a substrate as

$$D_T^{(1)} = D_0^{(1)} \exp\left(-\frac{E^*}{E}\right).$$ (4.12)

The value of E can be expressed in terms of J (the particle flux):

$$E = JStE_n(U),$$ (4.13)

where J is calculated from the equations presented in Reference 209. Substituting Equation 4.13 into Equation 4.12 and Equation 4.12 into Equation 4.8, we obtain

$$t_{tb} = \frac{V_1^{2/3}}{4\pi D_0^{(1)}} \exp\left(\frac{E^*}{JStE_n(U)}\right).$$ (4.14)

Knowing the time required for a temperature balance to be established and the irradiation time, we can rewrite Equation 4.5 with the above statements and conclusions in mind as

$$f(I,U,t_{ir}) = \left(\frac{t_{ir}}{t_{tb}}\right)^{3/2} \exp\left(1 - \frac{E_{Me-C_xH_y}}{E_n(U)}\right).$$ (4.15)

The exponent 3/2 in Equation 4.15 agrees with the law of heat (energy) dissipation in the environment. The 1 within the exponential function indicates that after a temperature balance is established, the process of complete dissociation of intermediate-layer molecules is exponential. Substituting Equation 4.15 into Equation 4.4, Equation 4.4 into Equation 4.3, and Equation 4.2 into Equation 4.3, we obtain the final expression for determining the total number of free bonds of organic molecules:

$$N_{sum} = N_0 \exp\left[-\exp\left(\frac{E_{Sub-C_xH_y}}{E_{Me-C_xH_y}}\right)\right] \times \left[1 + \left(\frac{t_{ir}}{t_{tb}}\right)^{3/2} \exp\left(1 - \frac{E_{Me-C_xH_y}}{E_n(U)}\right)\right].$$ (4.16)

From Equation 4.16, it follows that at $t_{ir} = 0$, Equation 4.16 changes to Equation 4.2 with the corresponding energies of individual bonds, $E_{Sub-C_xH_y}$ and $E_{Me-C_xH_y}$; at $t_{ir} < t_{tb}$, partial dissociation of contaminant molecules takes place, because the phase interface is deficient in input energy as a result of the energy being dissipated by the substrate and the substrate holder.

At $t_{ir} = t_{tb}$, complete irreversible dissociation is observed, and at $t_{ir} > t_{tb}$, free bonds in the layer of organic molecules cease to be formed, and all hydrocarbon molecule and atom bonds are broken to form chemical compounds with the material atoms of the phase interface.

Thus, after the ion–electron bombardment, the adhesion of the thin metal film to the substrate surface as a function of the process parameters is given by

$$A = N_0 \exp \left[-\exp \left(\frac{E_{Sub-C_xH_y}}{E_{Me-C_xH_y}} \right) \right]$$
$$\times \left[1 + \left(\frac{t_{ir}}{t_{tb}} \right)^{3/2} \exp \left(1 - \frac{E_{Me-C_xH_y}}{E_n(U)} \right) \right] \frac{1}{V_b N_a} \sum_{i=1}^{l} \Delta E_i, \qquad (4.17)$$

where V_b is the volume of the phase interface; N_a is the Avogadro constant; ΔE_i is the difference between the energies of individual bonds for the i component of the interface; and l is the number of interface components (materials).

The proposed analytical relationship will help reduce the volume of expensive and complex research into the effect that the parameters of ion–electron bombardment have on adhesion. The data derived from Equation 4.17 can be directly used in technology for fabricating DOEs and nano- and microelectronic elements.

4.3 EXPERIMENTAL INVESTIGATION INTO THE EFFECT OF ION–ELECTRON BOMBARDMENT PARAMETERS ON ADHESION

With the experimental and theoretical results in mind, and with a view to verifying whether the developed model agrees with the physical process, we have experimentally studied the relationship between the adhesion strength and the irradiation time (t_{ir}), the discharge current (I), and the accelerating voltage (U). The irradiation time determines the portion of energy supplied to the structure under study within the time interval t_{ir} at given values of I and U. The graph in Figure 4.7 suggests that over a period of $0 < t_{ir} \leq 3$ min, a temperature balance is being established in the metal–organic contamination–substrate system [247].

As the temperature balance is established, smaller portions of energy are being removed from the phase interface by the substrate and the holder, while the portion of the absorbed energy is increasing, thus intensifying the dissociation of organic molecules. At $t_{ir} \geq 3$ min, the energy supplied to the intermediate layer becomes quite sufficient for the complete dissociation of hydrocarbon molecules and formation of strong chemical compounds at the metal–substrate interface. Because of this, the irradiation time of $t_{ir} = 3$ min was taken for purposes of further studies. The resulting increase in the spread of adhesion-strength values apparently determines the formation of structurally different chemical compounds. But the bond strength between the metal film and substrate becomes equal to the substrate material's interatomic bonds.

This is evident from the comparative analysis of the substrate surface of the untreated and treated Me–C$_x$H$_y$–Sub structures once the metal film is removed (Figure 4.8a and b). To make the dielectric-surface modification more visual, the

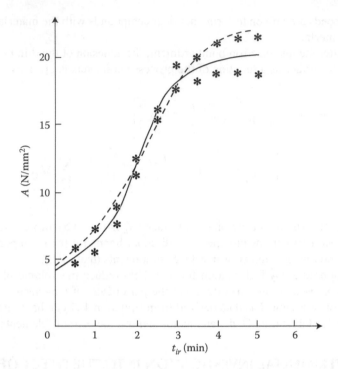

FIGURE 4.7 Relationship between copper film–polycrystalline substrate adhesion and the bombardment time of the Me–C_xH_y–Sub structure at $I = 100$ mA, $U = 2$ kV, and $t_{cont} = 10$ min. Relationship (4.17) is shown as a dashed line.

images in Figure 4.8a and b were obtained from the surface polished with the M20 powder, which provides a higher reactivity. As Figure 4.8c shows, within the limits of the disturbed layer, near-pit dislocation clusters are observed, indicating the formation of strong chemical bonds [96].

The adhesion–discharge current relationship (Figure 4.9) is nonlinear and comprises three sections: $0 \leq I \leq 40$ mA is a linear-growth section with a minimum value of ΔA and the absence of significant changes in the organic film's structure at the interface; $40 \leq I \leq 80$ mA is a section where exponential growth of A is observed and where the value of ΔA increases with increasing discharge current, reaching its maximum at $I = 80$ mA (intensive decomposition of the organic film and induction of broken interatomic bonds at the Me–C_xH_y–Sub interface take place); and $I \geq 80$ mA is a section where no adhesion growth is observed and where ΔA decreases: all molecules of the organic film are dissociated, forming better-ordered chemical compounds, and this makes any further increase in the current inexpedient.

The graph in Figure 4.10 (curves 1, 2, and 3) also provides support for the above analysis. Comparing the curves suggests that the maximum values of the adhesion strength (≈ 25 N · mm^{-2}) are reached when the Me–C_xH_y–Sub structure is treated with an ion–electron flux in the oxygen atmosphere.

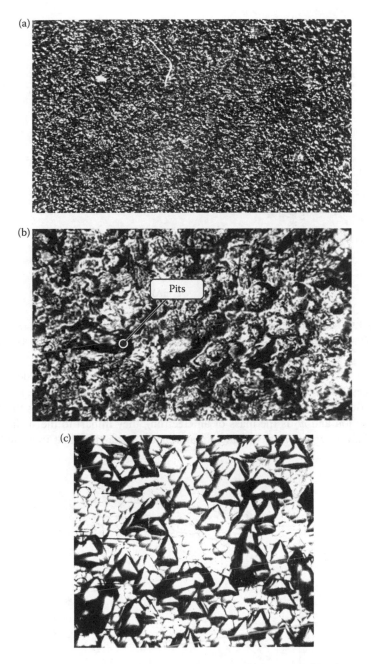

FIGURE 4.8 Appearance of the dielectric surface after the metal film is removed: (a) untreated structure (×36), (b) treated structure (×36), and (c) dislocation cluster seen on the dielectric substrate's surface after the metal film is removed from the treated structure (×400).

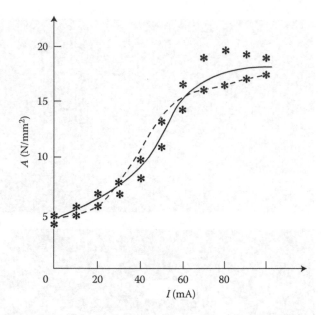

FIGURE 4.9 Copper film–polycrystalline substrate adhesion as a function of the discharge current: $U = 2$ kV, $t_{ir} = 3$ min, and $t_{cont} = 10$ min. Relationship (4.17) is shown as a dashed line.

In this case, it is the negative ions formed through mechanisms described in References 209 and 210 that mostly contribute to the interaction with the metal film's surface. Being much more massive than electrons ($m_i \gg m_e$, where m_i is the mass of a negative ion and m_e is the mass of an electron), they impart to the surface atoms of metal an energy several orders of magnitude greater than that imparted by the

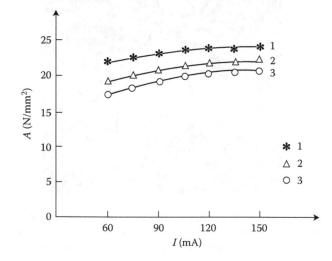

FIGURE 4.10 Copper film–polycrystalline substrate adhesion as a function of the discharge current: (1) oxygen, (2) air, and (3) argon. $U = 2$ kV, $t_{ir} = 3$ min, and $t_{cont} = 10$ min.

electrons. Thus, a much greater portion of the energy E_n (U) is also imparted to the atoms of the metal–chemical compound–substrate interface. This portion becomes sufficient to modify (redistribute) the bonds already established, forming modified chemical compounds characterized by greater binding energies.

Lower adhesion values in the air and argon atmospheres are due to a lower content of negative ions moving toward the sample. But to avoid the influence of surface bombardment by high-energy ions, an effect inherent in generating submicron structures, it is the argon atmosphere that needs to be used, as containing solely electrons. Because the parameters affecting the processes involved in the interaction between charged ion–electron beam particles and target atoms include the masses and velocities of the electrons and ions, the discharge current was reduced to 20 mA to study in detail how adhesion depends on the accelerating voltage. The curve in Figure 4.11 shows that the relationship $A = f(U)$ comprises three sections as well.

At $0 \leq U \leq 0.5$ kV, the distance over which the ion–electron beam propagates is less than the distance to the sample [209]. As a result, in the absence of surface bombardment, the organic film preserves its structure and the adhesion strength remains unchanged.

According to Reference 209, the etching of the material does not take place under these conditions. In the voltage range of $0.5 < U \leq 3.5$ kV, the hydrocarbon layer is modified. The significant steepness of the curve $A = f(U)$ at $1.2 \leq U \leq 3.5$ kV as compared with the interval $0.5 < U \leq 1.2$ kV is due to the larger portions of energy transferred from the particles of the ion–electron flux to the metal film's surface ($E_n[U]$) [209]. The $E_n(U)$ values required for the complete dissociation of the entire organic film are observed when the accelerating voltage is greater than 3.5 kV. The similar character of these relationships suggests that identical processes occur in the

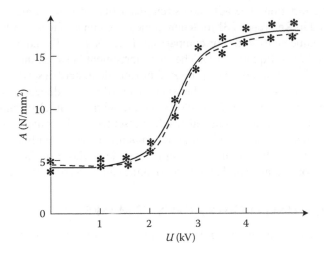

FIGURE 4.11 Copper film–polycrystalline substrate adhesion as a function of the accelerating voltage at $I = 20$ mA, $t_{ir} = 3$ min, and $t_{cont} = 10$ min. Relationship (4.17) is shown as a dashed line.

Me–C$_x$H$_y$–Sub structure under an ion–electron bombardment and that these processes obey the laws of heat conduction.

But the irradiation time and the accelerating voltage that determine the velocity of the charged particles when they hit the metal film's surface are the key factors in establishing the temperature balance in the Me–C$_x$H$_y$–Sub system and in the dissociation processes of the organic film's molecules, as suggested by a greater steepness of the related curves in Figures 4.7 and 4.11. In similar experiments with thicker films, the curve in Figure 4.7 was found to be extended along the axis of abscissas.

Thus, the relationships suggest that:

1. The adhesion of metal films to the surfaces of dielectric substrates during ion–electron bombardment is enhanced through the dissociation of organic molecules in the contamination layer at the metal–dielectric interface after a temperature balance in the Me–C$_x$H$_y$–Sub system has been established.
2. The maximum values of the adhesion strength are achieved at an irradiation time of at least 3 min, a discharge current of 80 mA, and an accelerating voltage of 4 kV.
3. The ion–electron bombardment of the Me–C$_x$H$_y$–Sub structure results in a 3.8–7.7-fold increase in the adhesion of metal films [248,249].

Under similar conditions, the ion–electron bombardment of a Me–Sub structure with the missing intermediate layer of organic contaminants, according to the mechanism proposed, results in the complete activation of metal–dielectric surface bonds at the interface. As a result of the activation, surface atoms of the materials in contact chemically interact along the entire metal–substrate interface. In this case, the adhesion increase is no less than 10-fold because of the greater density of dielectric leading to the greater number of free surface bonds than in the case of organic contamination. The adhesion strengths achieved here significantly surpass those reported in Reference 231 for technologically clean surfaces, while the irradiation time is reduced 10-fold. For comparison, Figures 4.7, 4.9, and 4.11 present the curves derived from Equation 4.17. The close agreement between the calculated and experimental data shows that the proposed model adequately describes the physical process of enhancing the adhesion between metal films and dielectric surfaces as a result of bombarding the Me–C$_x$H$_y$–Sub structure with an ion–electron flux.

The theoretical and experimental results presented in this chapter make it possible to relax the diffractive optics requirements for surface cleanliness that need to be satisfied for the fabrication of optical microreliefs whose profile aspect ratio is lower by a factor of tens than that of modern nano- and microelectronic components.

4.4 DEPOSITING HIGHLY ADHESIVE MASKS

The procedure for depositing highly adhesive masks consists of the following steps:

1. Repeat the sequences of operations described in Sections 2.6 and 3.8.
2. Apply a photoresist layer (by centrifuging at 3,000 r.p.m. for 20 s) in the case of applying a photoresist mask or deposit a 0.2-μm-thick chrome layer

on the substrate, using the vacuum deposition technique (in the case of applying a metallized mask).

3. Apply a photoresist layer (in the case of applying a metallized mask) through centrifuging at 3,000 r.p.m. for 20 s.

4. Dry the photoresist layer in the air for 30 min.

5. Dry the photoresist layer in the SPt 200 oven at 90°C for 30 min.

6. Expose the photoresist layer to ultraviolet light through a photomask for 70 s.

7. Develop the photoresist layer in a 0.3% solution of potassium hydroxide for 40–60 s.

8. Bake the photoresist layer in the SPt 200 oven at 100°C for 20 min and at 115°C for 40 min.

9. Etch the chrome layer in a selective etchant—a cerium solution consisting of $Ce(SO_4)_2$ (30 mL), H_2SO_4 (10 mL), and H_2O (200 mL)—until the chrome is completely removed.

10. Rinse the substrates in distilled water.

11. Dry the substrates at 90°C for 30 min.

12. Load the substrates into the working chamber.

13. Evacuate the working chamber to a pressure of $2 \cdot 10^{-2}$ Torr and then fill it with the process gas (argon) to a pressure of 1 Torr.

14. Repeat step 13.

15. Evacuate the working chamber to a pressure of $2 \cdot 10^{-2}$ Torr, activate the high-voltage gas discharge, and set the discharge current at 80 mA (controlled by admitting the process gas through a micro leak valve), the accelerating voltage at 4 kV, and the irradiation time at 3 min.

16. Deactivate the high-voltage gas discharge and cool the substrates under vacuum conditions for 10 min.

4.5 CHAPTER SUMMARY

The investigations discussed in this chapter showed what effect the physico-technical process parameters (irradiation time, discharge current, and accelerating voltage) have on the strength of adhesion between metal films and the surfaces of dielectric substrates. We have found that ion–electron bombardment of the metal–organic contamination–substrate structure stimulates the active dissociation of hydrocarbon molecules and that the adhesion strength between a metal film and the surface of a dielectric substrate increases as a result of bombarding the surface with an ion–electron flux.

The investigations provided the basis for an off-electrode plasma method for enhancing the adhesion of thin metal films by a factor of 4–7. The method has a low cost and energy consumption and achieves a 10-fold decrease in treatment time.

This chapter also offered some practical recommendations for using the proposed method in the technique for depositing highly adhesive masks as part of fabricating optical microstructures.

5 Etching the Surface Microreliefs of Optical Materials in Off-Electrode Plasma

Because optimal process regimes are difficult and expensive to determine experimentally, physico-mathematical methods for analyzing the parameters of plasma-chemical processes have found wide application.

Today, there are many models [108,190,250–252] describing plasma-chemical processes, but they either contain components that are difficult to ascertain experimentally (e.g., the K-layer [108,250] in the plasma-chemical etching model for polymers) or describe the mechanisms of plasma-chemical etching in general and thus do not allow any specific calculations to be carried out. Most models [253] involve significant simplifications to obtain analytical expressions, but this is not always acceptable.

Meanwhile, as the quality requirements for DOE microrelief profiles are becoming more demanding, a need arises for significantly increasing the etching anisotropy of substrate materials. Existing etching methods either fail to solve this problem or involve costly procedures. For that reason, diffractive optics developers tend to use new sources of low-temperature plasma whose properties meet the emerging requirements for DOE profiles. The requirements include [1] uniform microrelief etching across the entire substrate surface to a height of h_m, which is determined on a case-by-case basis from the equation $h_m = \lambda_w/(n-1)$, where λ_w is the wavelength and n is the material's refractive index; reducing the deviation of microrelief-step sidewalls from the vertical; and reducing the roughness of microrelief surfaces.

As Chapter 1 showed, a convenient low-temperature plasma source for fabricating diffraction microreliefs is a high-voltage gas-discharge device. The device is simple and generates a directed CAP flux with plasma particles distributed across its cross section in a highly uniform manner. But the literature is silent on etching modeling and the mechanisms of plasma-chemical and ion-chemical surface etching in high-voltage gas-discharge plasma. For that reason, this chapter addresses the problems of theoretical and experimental research into the mechanisms of plasma-chemical and ion-chemical surface etching as well as the problems related to developing models and methods for surface etching with off-electrode plasma.

5.1 PREPARING SAMPLES FOR AN EXPERIMENT IN ETCHING THE SURFACE MICRORELIEFS OF OPTICAL MATERIALS IN OFF-ELECTRODE PLASMA

The use of the new reactor, which uses high-voltage gas discharge to generate low-temperature plasma outside the electrode gap, entailed many complex theoretical and experimental investigations into surface-etching mechanisms. In these investigations, we used silicon dioxide as a primary material. To etch an optical microrelief with off-electrode plasma, we also used silicon carbide substrates and diamond-like films. The fabrication of diffraction microreliefs comprised the following operations: preparing 30×20 mm^2 substrate surfaces for applying a 0.2-μm chrome layer or a 0.5–1 μm photoresist layer; applying a 0.5–1 μm photoresist layer after applying a chrome layer; using the standard photolithography techniques to form microrelief masks with periods of $T = 12$–32 μm; and treating the masks with high-voltage gas-discharge plasma to enhance their adhesion strength.

The preparation of the substrate surfaces consisted of cleaning the surfaces in two steps: rough cleaning and final cleaning.

Rough cleaning. Rough cleaning comprised the following steps:

1. Rubbing the substrates with a cambric cloth soaked in ethyl alcohol on both sides, to remove large mechanical particles
2. Boiling the substrates in distilled water for 10 min
3. Boiling the substrates in an alkali solution for 10–15 min
4. Boiling the substrates in distilled water for 10 min
5. Boiling the substrates in ethyl alcohol for 10 min
6. Drying the substrates in the SPt 200 oven for 10–15 min at 120–130°C
7. Cooling the substrates in the desiccator

This procedure removes both organic and mineral contaminants.

Final cleaning. Immediately before the masks were deposited, final cleaning was accomplished as described in Chapter 3 with off-electrode plasma at a discharge current of 3 mA, an accelerating voltage of 1.2 kV, and an exposure time of 10 s. The process gas was air. The proposed cleaning technique helped attain a level of 10^{-9} g/cm^2, a level of highly precise surface cleaning.

Cleanliness measurement. The degree of surface cleanliness was measured with the tribometric method [254,255] and the wettability method discussed in detail in Chapter 2.

Evaporation. Chrome layers were deposited on the substrate surfaces in UVN-2M-1 with the thermal vacuum evaporation technique [232,256,257].

Photoresist application. Photoresist layers were deposited through centrifuging by using the standard technique [108]; FP-1318 photoresist was used, and it yielded 0.5–1 μm layers as a result of centrifuging at 3,000 r.p.m. for 20 s.

Photoresist drying. The photoresist layers were dried in two steps: first, in the air for 30 min and then in the SPt 200 oven at 90°C for 30 min.

Light exposure. The photoresist layers were exposed to ultraviolet light through photomasks for 70 s.

Photoresist development. The photoresist layers were developed in a 0.3% solution of potassium hydroxide for 40–60 s.

Baking. The photoresist layers were baked in the SPt 200 oven at 100°C for 20 min and at 115°C for 40 min.

Etching. The chrome layers were etched in a selective etchant—a cerium solution consisting of $Ce(SO_4)_2$ (30 mL), H_2SO_4 (10 mL), and H_2O (200 mL)—until the chrome was completely removed.

Adhesion enhancement. Mask adhesion was enhanced through treatment with off-electrode plasma at a discharge current of 80 mA, an accelerating voltage of 4 kV, and an exposure time of 3 min.

To investigate the formation mechanism of optical microreliefs observed in the catalytic-mask method, similarly prepared samples on silicon substrates were used. The only two differences were that, instead of chrome, aluminum was deposited as the first masking sublayer and that the photoresist layer was completely removed after etching the aluminum layer.

During and after the plasma treatment, the samples' parameters were measured with several instruments. Microrelief parameters were measured with a 170311 profilograph/profilometer; an MII-4 optical microinterferometer; P4-SPM-MDT and SMENA scanning probe microscopes in combination with NT MDT Solver P47H and Solver PRO-M; a Zygo NewView 5000 surface profilometer; a Philips XL40 scanning electron microscope; and an FEI FIB-200 focused ion beam microscope. The composition of the material deposited on the cathode surface during etching was studied through x-ray structure analysis with a DRON-2.0 x-ray diffractometer. Resistivity was measured with a TsIUS-2 unit. The conductivity types of the layers in the structure under study were determined from the value and sign of the thermal electromotive force (EMF). And trends of how Si atoms were distributed in aluminum were determined through metallographic analysis of the samples' cross sections.

Given that the parameters were measured with standard precision equipment, the results we obtained can be considered reliable.

5.2 MECHANISMS OF PLASMA-CHEMICAL AND ION-CHEMICAL SURFACE ETCHING

Plasma-chemical etching with high-voltage gas-discharge plasma involves ion bombardment directed along the normal to the sample surface. This enhances the anisotropy of the process and the surface etching as a result of reactive particles (such as atomic fluorine) forming directly at the substrate surface. Their formation results from ion–molecule interactions between negative ions and neutral process-gas molecules that were initially adsorbed from the gas phase. Ion bombardment during plasma-chemical etching with off-electrode plasma is a primary source of CAPs. We will illustrate this by looking into the reactions that occur in this plasma and cause CAPs (neutral radicals) to form, assuming by way of illustration that the plasma-forming gas is CF_4.

Traditionally, CAPs in halocarbon-14 plasma generated by HF and SHF discharges have been known to form both in the plasma bulk and at the substrate surface as a result of neutral molecules being dissociated by an electron impact. With the results presented in Reference 258 serving as the basis, these processes can be described as the following reactions:

$$e^- + CF_4 \rightarrow CF_3^+ + F^* + 2e^-; \tag{5.1}$$

$$e^- + CF_4 \rightarrow CF_3^+ + F^* + e^-; \tag{5.2}$$

$$e^- + CF_4 \rightarrow CF_3^* + F^-, \tag{5.3}$$

where e^- stands for electrons; F^* for fluorine radicals; F^- for negative fluorine ions; C for carbon atoms; and F for fluorine atoms.

As noted in Chapter 1, charges in off-electrode plasma are strictly divided according to their direction. At the moment that a negative ion or electron is formed that will move toward the substrate, a positive ion must be formed that will move toward the cathode [256]. On the other hand, interactions resulting in two or more negative particles are possible in high-voltage gas discharge, but simultaneously with these interactions, other reactions must take place to form enough positive ions; that is, the balance of moving charged particles must be preserved. Where this condition (the inequality $\gamma Q \geq 1$) is violated, the discharge ceases. This may happen when the energy of negatively charged particles (ions or electrons) is insufficient for a positive ion to be formed during interaction with process-gas molecules—for example, at a distance of d_{max} from the electrodes of the gas-discharge device. From this viewpoint, the first reaction (electron-impact dissociative ionization) is the most acceptable.

Reference 209 stresses that electron attachment to neutral atoms is the main mechanism that causes electrons to lose energy at an electrode voltage of 0.5–2 kV, a range that includes the region of plasma-chemical etching. In this case, CAPs are short-lived [94,95]. Upon the extinction of a F* radical, an electron joins it to form a negative ion, F^-:

$$F^* + e^- + e^- \rightarrow F^- + e^-. \tag{5.4}$$

Because off-electrode plasma is generated as a directed flux, the probability of it colliding with the reactor's walls is low, so the effect that the heterogeneous reactions of charged particles' recombination on the working chamber's walls have on the processes proceeding in the plasma is negligible. For electron–ion recombination to take place, it is not only necessary that the plasma contain enough free electrons but also that their energies be lower than the ion ionization energy. In the case under analysis, these conditions are not met; therefore, the primary mechanism of charge loss in plasma-chemical etching with off-electrode plasma is ion–ion recombination [94]. We will not consider the excitation and ionization of process-gas molecules by an electron impact, since both take place at a higher pressure [94]. Thus, the following are the primary reactions that are possible in the bulk of off-electrode plasma and that agree with the above considerations [210]:

$$e^- + CF_4 \rightarrow CF_3^+ + F^- + e^-; \tag{5.5}$$

$$F^- + CF_4 \rightarrow CF_3^+ + 2F^-; \tag{5.6}$$

$$F^- + CF_3^+ \rightarrow CF_4. \tag{5.7}$$

The ionization of a process-gas molecule by a F^- ion (5.6) is valid because, according to Equation 3.7, the negative ion's energy exceeds the ionization energy of the CF_4 molecule along the entire path to the substrate. The F^- ion's energy in this case changes from 400 eV after the first collision to 100 eV or below before the ion collides with a molecule adsorbed at the substrate surface. In the latter case, some of the ion's energy (on the order of ionization energy) is spent to ionize the CF_4 molecule and the remainder to destroy or weaken the bonds of surface atoms of the material being treated. According to Reference 210, this interaction occurs in the following manner:

$$F^- + CF_4 + S_p \rightarrow CF_3^+ + 2F^* + 2e^-, \tag{5.8}$$

where S_p is a surface element. The F* radicals being formed do not have time to join an electron and instead immediately react with the surface atoms to form volatile compounds of the form

$$nF^* + A \rightarrow AF_n \uparrow, \tag{5.9}$$

where A is an atom of the material being treated and n is the number of fluorine atoms required to remove one A atom. Thus, the F^- ions can be said to provide a targeted delivery of F* radicals to the surface being treated—that is, for each F^- ion in the plasma bulk, there is a corresponding F* radical at the surface. The reaction products are removed from the working chamber by evacuation facilities.

When the F^- ion's energy becomes equal to or less than the CF_3^+ ion's ionization energy, the two recombine according to Equation 5.7 to form a process-gas (CF_4) molecule.

Thus, from reactions (5.5) through (5.9), it follows that:

1. In off-electrode plasma, CAPs (F* radicals) are formed directly at the surface of the material being treated. This prevents their termination in interaction with other plasma-forming particles.
2. F$^-$ ions play the key role in generating CAPs.
3. One collision of a F$^-$ ion and a process-gas molecule adsorbed at the surface forms two CAPs, and the surface acts as a catalyst transforming the F$^-$ ions (formed after the CF_4 molecule is ionized) into neutral F* radicals.

4. CF_3^+ ions moving toward the cathode (see Figure 3.11) do not participate in forming $(C_xF_y)_n$ polymers at the surface being treated, and this in itself rules out the possibility of the surface being coated with carbon-bearing compounds and other passivating particles.

These features definitely improve the efficiency of plasma-chemical etching.

The etching mechanism for the ion-chemical region of off-electrode plasma is somewhat different from that discussed above. The principal difference is that the intense bombardment and heating of the surface by high-energy F^- ions (at 100–500 eV) impede the adsorption of process-gas molecules at the surface, thereby resulting in the low probability of CAPs being involved in the etching [212]. The material is removed through both F^- ion-induced physical sputtering and the subsequent chemical interaction of the ions with the material's atoms. The chemical reactions in this case are similar to those described for plasma-chemical etching. As the ion-chemical etching mechanism is adequately described in Reference 190, we will not go into further details here. Of much greater interest are the processes of plasma-chemical and ion-chemical etching occurring in high-voltage gas-discharge CF_4 plasma when oxygen is added to it. In this case, by analogy with reactions (5.5) through (5.7) and in addition to them, other types of reaction appear to be possible in CF_4/O_2 plasma [210]—namely:

Ionization:

$$e^- + O_2 \rightarrow O^+ + O^- + e^-; \tag{5.10}$$

$$e^- + OF \rightarrow O^+ + F^- + e^-; \tag{5.11}$$

$$F^- + O_2 \rightarrow O^+ + O^- + F^-; \tag{5.12}$$

$$F^- + OF \rightarrow O^+ + 2F^-; \tag{5.13}$$

$$O^- + CF_4 \rightarrow CF_3^+ + F^- + O^-, \tag{5.14}$$

Recombination:

$$O^+ + F^- \rightarrow OF; \tag{5.15}$$

$$CF_3^+ + O^- \rightarrow COF_2 \uparrow + F^*. \tag{5.16}$$

The literature relating to this topic [260] observes that the volatile compound of carbonyl difluoride (COF_2) decomposes to release free fluorine and form carbon monoxide:

$$COF_2 \rightarrow CO \uparrow + 2F^*. \tag{5.17}$$

The radicals formed in reactions (5.16) and (5.17) and electrons combine according to Equation 5.4 to become F⁻ ions that, according to Equation 5.8, undergo a reverse change at the surface during plasma-chemical etching. Such changes should be expected as well from O⁻ ions. Oxygen radicals, being highly chemically active, compete with F* radicals for the active sites at the surface. At a certain O_2 concentration in plasma, this can decrease the plasma-chemical etch rate.

5.3 ETCHING MODEL: PRIMARY EXPRESSIONS; ALGORITHM AND SOFTWARE FOR CALCULATING THE ETCH RATE

We will calculate the flux of negative ions incident on the substrate by using the mechanism of ion–electron emission and the theory of avalanche breakdown within the gas gap. In this case, the density of the current in the cathode area can be expressed as

$$j = \frac{I}{qS_K}, \tag{5.18}$$

where I is the discharge-circuit current (A); S_K is the cathode's surface area (cm²); and q is the geometric transparency of the gauze anode. This value can also be derived from [261]

$$j = j_i^+(1+\gamma_e), \tag{5.19}$$

where j_i^+ is the current density of positive ions incident on the cathode (A/cm²) and γ_e is the secondary-emission coefficient. By solving the system of Equations 5.18 and 5.19, we obtain the density of the ion current in the cathode area:

$$j_i^+ = \frac{I}{qS_K(1+\gamma_e)}. \tag{5.20}$$

According to the theory of one-electron approximation, the concentration of multicharged ions is, to a first approximation, negligible, and so the flux of positive ions incident on the cathode is equal to

$$J_i^+ = \frac{I}{eqS_K(1+\gamma_e)}, \tag{5.21}$$

where e is the electron charge (C).

Bombarding the cathode with positive ions causes electron emission whose flux is equal to [94]

$$J_e^K = J_i^+\gamma_e. \tag{5.22}$$

Thus, substituting Equation 5.21 into Equation 5.22, we obtain

$$J_e^K = \frac{I\,\gamma_e}{eqS_K(1+\gamma_e)}. \tag{5.23}$$

When the discharge is governed by electron attachment, the theory of avalanche breakdown determines an electron flux incident on the anode as [94]

$$J_e^A = J_e^K \exp[(\alpha - \alpha_n)d_{\max}], \tag{5.24}$$

where α is the ionization coefficient and α_n is the attachment coefficient. That electron attachment is the main mechanism that leads to electron loss in the case under analysis can also be confirmed by calculating the electron-current density in the anode area (j_e) from Equation 5.24 without taking account of α_n, and by then comparing j_e with the registered anode-current density (j). The comparison shows that the calculated electron flux does not match the discharge-circuit current: $j_e \gg j$. This can be explained by the presence in the working chamber of the electronegative fluorinated gas, in which electrons intensely attach to neutral atoms and molecules [262–264]. Attachment acts as a factor that reduces the ionization rate of process-gas molecules: in this case, electron multiplication in the avalanche depends on the difference between the coefficients of ionization and attachment, and Equation 5.24 becomes determinant. The only drawback with this equation is that the difference $\alpha - \alpha_n$ is quite difficult to determine. This monograph proposes that this problem be solved with the equation [261]

$$\frac{J}{J_e^K} = \frac{\alpha}{\alpha - \alpha_n}\exp[(\alpha - \alpha_n)d_{\max}] - \frac{\alpha_n}{\alpha - \alpha_n}, \tag{5.25}$$

where J is the full flux of negatively charged particles that have reached the anode (the registered flux). Equation 5.25 cannot be solved for $\alpha - \alpha_n$ with standard techniques. To overcome this difficulty, we used the line-bisection technique [265] from a MATLAB® software package. The technique made it much easier to determine numerical data for the model. By substituting the difference $\alpha - \alpha_n$ obtained from that technique into Equation 5.24, it is easily possible to calculate J_e^A.

A feature of high-voltage gas discharge is the electron-focusing effect [266]. From this, it follows that the electron flux from the cathode surface will be greater than the electron flux incident on the sample by the focusing value. Now, to determine the electron flux incident on the substrate surface, we should multiply J_e^A by the focusing coefficient of the electron flux, η. Substituting Equation 5.23 into Equation 5.24 and taking into account the equation $\eta\,J_e^A = J_e$, we finally obtain

$$J_e = \frac{I\gamma_e\eta}{eqS_K(1+\gamma_e)}\exp[(\alpha - \alpha_n)d_{\max}]. \tag{5.26}$$

As the anode does not emit any particles, the total charged particle flux incident on the substrate surface is determined by the fluxes of electrons and negative ions as

$$J = J_e + J_i^-. \tag{5.27}$$

Substituting Equation 5.26 into Equation 5.27 and expressing J in terms of the discharge current, we obtain the flux of negative ions incident on the substrate surface:

$$J_i^- = \left(1 - \frac{d}{d_{max}}\right) \frac{I}{qeS_K} \left(1 - \frac{\gamma_e \eta}{(1+\gamma_e)} \exp[(\alpha - \alpha_n)d_{max}]\right), \tag{5.28}$$

where $(1 - d/d_{max})$ is a multiplier that numerically characterizes those ions from the aggregate flux that reach the sample surface and that contribute to its etching at a given value of d_{max}.

To calculate the CAP flux incident on the sample surface, we will use the theory of dissociation under ion bombardment. The theory holds that the CAPs that provide etching are formed directly at the surface of the material being treated as a result of ion–molecule interactions between ions and neutral molecules [190]. Thus, the flux of adsorbed and undissociated process-gas molecules incident on the substrate surface is given by

$$J_n = 1.074 \times 10^{17} \frac{p}{\sqrt{MT}}, \tag{5.29}$$

where p is the pressure (Pa) and T is the temperature of the surface being treated. We propose that the effect of the temperature on the molecule adsorption under high-voltage gas discharge be expressed as

$$T = T_0 \left(1 + \exp\left(\frac{E_n - E_n^{gd}}{kT}\right)\right), \tag{5.30}$$

where T_0 is the ambient temperature (K); $E_n^{gd} = 53.445$ is the ion's energy (eV) that is observed at the moment the ion approaches the surface and that corresponds to the gas-discharge device's electrode voltage at which condition (3.9) is satisfied and at which the etching process reaches an optimum; E_n is the same energy, except that it corresponds to the operating voltage at the electrodes; and k is the Boltzmann constant (J/K).

Dissociation of each such molecule is held to form one CAP. But in reality, not all CAPs formed through dissociation can react with the material being treated because some of them can be desorbed from its surface under the action of ions. And the number of CAPs depends on the number of ions themselves. Given these factors, the CAP flux incident on the substrate surface and directly etching it is equal to

$$J_a = J_i^- \theta, \tag{5.31}$$

where θ is the degree of surface filling by active particles.

Note that we used the theory of dissociation because it allows calculations for CAPs that are directly at the surface being treated, as opposed to those in the plasma bulk. This appreciably improves the model's accuracy and provides a more realistic picture of etching processes.

Consider the etching process that takes place in the plasma-chemical region of off-electrode plasma and that is determined by the particle fluxes J_i^-, J_n, and J_a. Because the processes of etching and desorption under the action of electrons are negligible, the expression for θ can be written as follows [267] if the plasma does not contain passivating particles:

$$\theta = \frac{1}{\left[1 + ((k_1 + k_2)J_i^- / (s_a J_n))\right]}, \tag{5.32}$$

where s_a is the coefficient of CAP attachment to the surface. The unknown desorption coefficient in Equation 5.32 can be derived from Equation 3.7 and the known expression [214]

$$k_2 = \frac{3}{4\pi^2} \beta \frac{4mM}{(m+M)^2} \frac{E_n}{E_b}, \tag{5.33}$$

where β is a unitless parameter depending on the mass ratio M/m of the incident ion (m) and the target particle (M), and E_b is the binding energy of the target particle and the surface (eV).

Thus, the etch rate for the plasma-chemical region can be expressed as

$$V_{pce} = \frac{Bk_1 M_{mat}}{\rho N_A} J_i^- \theta, \tag{5.34}$$

where B stands for values of the penalty function that are obtained from a natural experiment and that are constants derived from the relationship $B = f(I)$; M_{mat} is the molecular mass of the material being etched (g/mol); ρ is the material's density (g/cm^3); and N_A is the Avogadro constant (1/mol). The numerical calculation of the flux J_i^- with Equation 5.28 shows that in the case of electron attachment the parenthetical term characterizing the flux J_e is negligible. Thus, recalling Equation 3.8, we can finally express Equation 5.34 as

$$V_{pce} = \frac{Bk_1 M_{mat}}{\rho N_A} \left(1 - \frac{d}{2ctg[\pi(U - U_1)/2U]}\right) \frac{I}{qeS_K} \theta. \tag{5.35}$$

From Equation 5.35, it follows that if the denominator characterizing the value of d_{max} in the parenthetical fraction is less than or equal to the distance to the substrate, etching is impossible. This agrees with condition (3.9). On the other hand, if the sample's surface temperature exceeds the maximum allowable temperature at which the adsorption of CAPs at the surface is still possible, then the etch rate equals zero as well, and this does not contradict [212]. Both phenomena were also observed experimentally, thereby confirming the correctness of the selected etching mechanisms.

The etch rate for the ion-chemical region is given by

$$V_{ich} = \frac{Bk_{1,3}M_{mat}}{\rho N_A} J_i^-, \tag{5.36}$$

where $k_{1,3}$ is the coefficient of the aggregate effect of physical sputtering and chemical reactions on the etching process. Since many studies [252,268,269] stress that these etching mechanisms are not additive and that it is incorrect to restrict oneself to determining their sum because the mechanisms stimulate each other, it is proposed that the sum be multiplied by a multiplier that links the mechanisms. This is an original solution that agrees with the purpose of this monograph.

In the case of high-voltage gas discharge, we propose that the coefficient $k_{1,3}$ be expressed as

$$k_{1,3} = (k_1 + k_3) \left| \exp\left(\frac{U - U_{gd}}{U} \right) - 1 \right|. \tag{5.37}$$

Expression (5.37) is valid because a decrease in U increases the contribution of the chemical component (k_1) to the etching process and decreases that of the physical component (k_3), while the reverse is true if the electrode voltage is increased. This agrees with established opinion [190,252]. Recalling Equations 5.28 and 5.37, we can express V_{ich} as

$$V_{ich} = \frac{B(k_1 + k_3)M_{mat}}{\rho N_A} \left(1 - \frac{d}{2ctg[\pi(U - U_1)/2U]} \right)$$
$$\times \frac{I}{qeS_K} \left| \exp\left(\frac{U - U_{gd}}{U} \right) - 1 \right|. \tag{5.38}$$

The coefficient k_3 in Equation 5.38 is calculated similarly to k_2 with the difference that the target particles in this case are the surface atoms of the material being treated that have a binding energy typical of these atoms.

Obtaining actual quantitative etch-rate values from the models developed is difficult and time-consuming [270–272]. This impedes the use of the models for adjusting the etching conditions directly during etching. To solve this problem, we propose using a special software application based on Microsoft's Visual Basic 5.0. The application automatically calculates the quantitative values of etch rates and etching parameters in real time. The application comprises three core program modules used to complete etching-related tasks in automatic or semiautomatic modes. These modules calculate the rates of plasma-chemical etching and ion-chemical etching and the secondary-emission coefficient.

The application also includes databases for storing parameter values for consumables—process gases and cathode materials. To increase the calculation rate, the data exchange between the modules and databases runs automatically.

To obtain the numerical values of the plasma-chemical etch rate, the application uses expressions (5.32) and (5.35). The values of those parameters that are used in Equations 5.32 and 5.35 but that are not functionally related, such as electrode voltage, chamber pressure (p), surface temperature (T), and electron-focusing coefficient (η), are saved to a dedicated file in the database in which each value of U has its own values of p, T, and η for each value of the discharge current. The entered values of U must fall within the range of $0 < U \leq 1{,}000$ V, the region of plasma-chemical etching. The etch rate is calculated for both the U values entered in the database and the intermediate values, and the linear-interpolation method is used for the calculations. After the input data are entered (Figure 5.1) and the process gas selected by loading its parameters (Figure 5.2) from the database, the condition $d_{max} > d$ is verified.

If this condition is not satisfied, no further calculation takes place and the zero value of the etch rate is displayed. Otherwise, the quantitative calculation continues that uses Equation 3.7 to determine the energy that the negative ion has after n collisions in approaching the surface. The value of n varies from one to the number of collisions that the particle has time to undergo when covering the distance (d) from the cathode to the sample, and the number is derived from the equation $n' = dec(d/\lambda)$, where dec is a rounding-down operator. Thus, each value of U has its own n used for calculating ΔU_n.

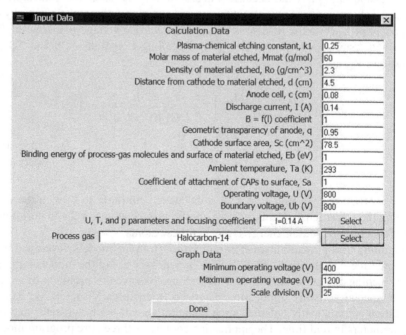

FIGURE 5.1 Input-data window for calculating the plasma-chemical etch rate (in this figure and in Figure 5.4, the symbol ^ stands for "raised to the power of"; for example, the g/cm^3 in the field Density of material etched, R_0, (g/cm^3) means g/cm³).

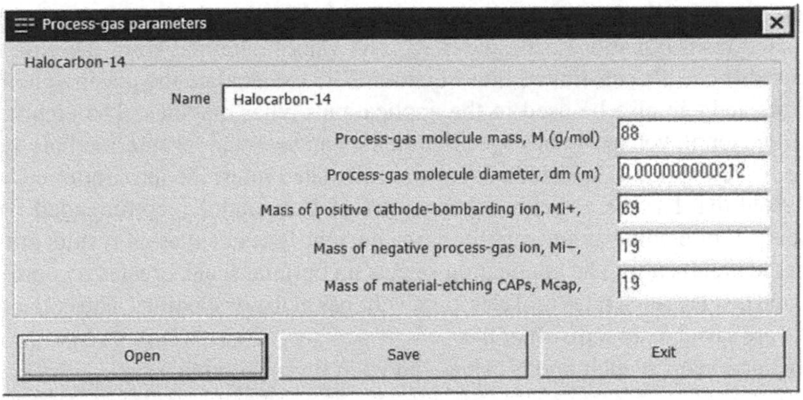

Process-gas parameters ☒

Halocarbon-14

Name	Halocarbon-14

Process-gas molecule mass, M (g/mol)	88
Process-gas molecule diameter, dm (m)	0.000000000212
Mass of positive cathode-bombarding ion, Mi+,	69
Mass of negative process-gas ion, Mi−,	19
Mass of material-etching CAPs, Mcap,	19

Open	Save	Exit

FIGURE 5.2 Database window for entering process-gas parameters.

The calculations that use Equations 3.6 and 3.7 continue until n reaches the value of n'. The initial energy of a charged particle depends on the voltage supplied to the gas-discharge device's electrodes—that is, $E_0 = eU$. The results obtained from Equations 3.6 and 3.7 are used for calculating the desorption coefficient of CAPs (Equation 5.33). The application automatically quantifies the β parameter. This function is accomplished through the use of a dedicated database that stores the values of the M'/m ratio and the related values of β.

The graph of the function $β = f(M'/m)$ [273] (see Figure 5.3) illustrates the relationship between these values with a scale that allows M'/m values to be entered in increments of $h = 0.05$. The values of β corresponding to the intermediate values of M'/m are calculated from linear-interpolation equations.

Because of electron attachment, off-electrode plasma contains few free electrons during plasma-chemical etching, and the ionization coefficient (α) equals the attachment coefficient ($α_n$). As a result, the application calculates the negative-ion flux incident on the substrate from the simplified expression

$$J_i^- = \left(1 - \frac{d}{2ctg[\pi(U - U_1)/2U]}\right)\frac{I}{qeS_K}. \tag{5.39}$$

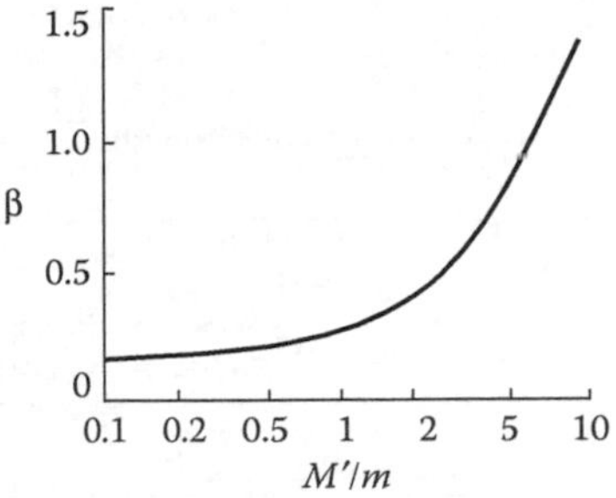

FIGURE 5.3 Relationship between β and the M'/m ratio.

The flux of neutral process-gas molecules incident on the substrate surface is calculated from Equations 5.29 and 5.30. The values obtained from the equations discussed above are substituted into Equation 5.35 to calculate the plasma-chemical etch rate and can also be used in the application's other modules. The etch rate is quantified for all voltage values specified in the database and for intermediate values (if the related option is selected). From the calculated rates, the maximum etch rate is selected that has the corresponding values of U, p, and T recommended by the application for etching at a given discharge current. The calculation results are also presented graphically. The application selects an optimal scale, creates a coordinate grid, and displays etch rates and electrode voltages at the grid points. The calculation results are saved to an individual file.

The ion-chemical etch rate is calculated from the expression

$$V_{ich} = \left(\frac{B(k_1 + k_3)M_{mat}}{\rho N_A} \right) \left(1 - \frac{d}{2ctg[\pi(U - U_1)/2U]} \right)$$
$$\times \frac{I}{qeS_K} \left| \exp\left(\frac{U - U_{gd}}{U} \right) - 1 \right| \times \left(1 - \left(\frac{\gamma_e \eta}{1 + \gamma_e} \right) \exp[(\alpha - \alpha_n)d_{max}] \right). \quad (5.40)$$

This expression shows that in this case the procedure is mostly similar to that described earlier for calculating the plasma-chemical etch rate. The only difference is that the input data include the cathode material (see Figure 5.4).

The required file is loaded from the database. The voltage values stored in the database are adjusted to fall within the range $1,200 \leq U \leq 3,500$ V, the ion-chemical region, and then the secondary-emission coefficient is calculated. The coefficient k_3

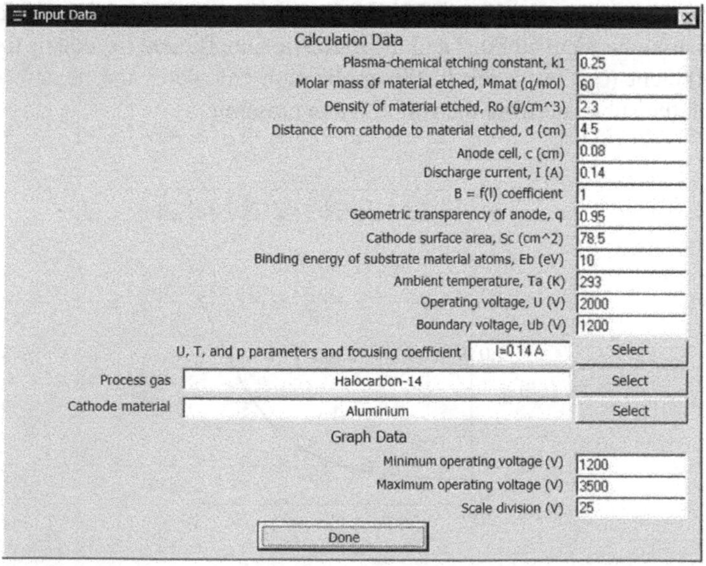

FIGURE 5.4 Input-data window for calculating the ion-chemical etch rate.

is calculated similarly to k_2 with the difference that now the target particles are the substrate's surface atoms that have a binding energy typical of them. As noted earlier, during ion-chemical etching, the free electrons in off-electrode plasma are formed at electrode voltages of $U > 2,000$ V. This enables us to use Equation 5.39 to calculate the negative-ion flux incident on the substrate in the range $1,200 < U < 2,000$ V (or $U_{gd} < U < 2,000$ V). At $U > 2,000$ V, secondary emission and the electron-focusing coefficient must be taken into account [84]. As a result, the expression for quantifying J_i^- changes to Equation 5.28. The unknown parameter in Equation 5.28 is γ_e, which is calculated by the appropriate module (Figure 5.5) from the expressions [96]

$$\gamma_e = \frac{\pi \xi_2}{4} \frac{Z(3d_p)^4}{d_p^3 a^2} \left(\frac{U_p}{m_e}\right)^{1/2}$$

$$\times \left[\frac{\exp(-U_p/kT_{eff})}{U_p/kT_{eff}} + E_f(U_p/kT_{eff})\right], \tag{5.41}$$

$$T_{eff} = \frac{1}{27\pi^{3/2}\xi_0 S} \frac{4M_i M^{me} eU}{k(M_i + M^{me})^2}, \tag{5.42}$$

where m_e is the electron's mass; U_p is the work function of the metal; a^2 is the ratio of the thermal conductivity coefficient to the thermal conductivity of the metal's unit volume; T_{eff} is the efficient temperature where the positive ion hits the metal; k is the Boltzmann constant, $\xi_2 = 0.99$ [96]; S is the number of degrees of freedom of a lattice

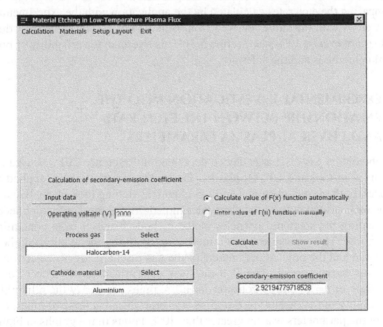

FIGURE 5.5 Window for calculating the secondary-emission coefficient.

site, $\xi_0 = 1.486$ [96]; M_i is the mass of a positive ion bombarding the cathode surface; and Mme is the mass of an atom of cathode material. These unwieldy expressions can be simplified by introducing the function

$$F(X) = \frac{\exp(-X)}{X} + E_f(X),$$

where $X = U_p/kT_{eff}$ and $E_f(X) = -\int_{-X}^{\infty} \exp(-U)/U \, dU$ is an exponential integral function.

With this software application, the function $F(X)$ can be calculated both automatically and manually for each value of U for the process gas and cathode material selected from the database. In the manual mode, a table is displayed that lists values for the function $F(X)$, each of which has a corresponding value of X. After the user selects the nearest value of the function $F(X)$ available for the X already calculated, the application displays the secondary-emission coefficient.

This value of γ_e is substituted into Equation 5.28 to calculate J_i^- as well as the ionization and attachment coefficients because the appearance of free electrons in the plasma violates the equality $\alpha = \alpha_n$, thereby making it necessary to know the numerical values of the coefficients.

The coefficients α and α_n are quantified by means of expression (5.25) with the known left part. That expression can be solved for $(\alpha - \alpha_n)$ with the line-bisection technique [265]. The difference $(\alpha - \alpha_n)$ thus obtained is used in Equation 5.40 to calculate the ion-chemical etch rate. The calculation results are presented similarly to those for plasma-chemical etching (see Figure 5.6).

Comparing the etch rates calculated by the application with the experimental values as well as with the values derived from Equation 5.40 shows that their discrepancy does not exceed 10% (see Section 5.3). This attests to the reliability of both the evaluation methods and the software.

5.4 EXPERIMENTAL INVESTIGATION INTO THE RELATIONSHIP BETWEEN THE ETCH RATE AND PHYSICAL PLASMA PARAMETERS

The high-voltage gas-discharge device described in Reference 210 was used to etch the samples in CF_4 and CF_4/O_2 plasma. Depending on the voltage applied to the electrodes, both plasma-chemical and ion-chemical etching took place. Because the samples were placed outside the electrodes, no additional measures were necessary for preventing the sputtered particles of the cathode material from contaminating their surfaces. The experiment was aimed at determining the relationship between the etch rate and the process parameters of the gas-discharge device: electrode voltage and discharge current. The etch rate was determined as the ratio of depth to an etch time of $t = 10$ min, a constant for all regimes. To verify the reliability and reproducibility of the results, each regime was reproduced at least 10 times, and the spread of the parameters was no greater than 10%. Points in the graphs in Figure 5.7 represent the mean of the etch rate for measurements taken on 10 samples. The figure

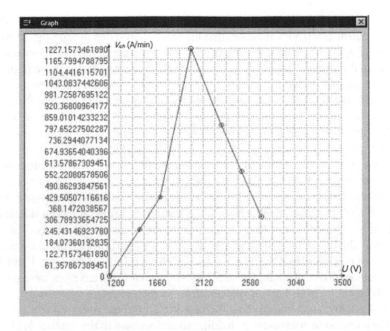

FIGURE 5.6 Relationship between the ion-chemical etch rate and the electrode voltage of the gas-discharge device (calculated by the application for $I = 140$ mA).

shows the experimental relationship between the etch rate and the electrode voltage for different discharge currents.

The curves show that at a low voltage (0.5–1 kV) and a high current (120–140 mA; curves 1 and 2), plasma-chemical etching takes place.

The increasing etch rate in the voltage range 0.5–0.8 kV indicates that an increase in the voltage leads to an increase in the rate at which CAPs are formed with the assistance of negative ions with energies (on the order of 50 eV if derived from Equation 3.7) that

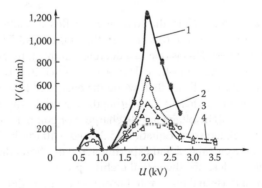

FIGURE 5.7 Relationship between the etch rate in CF_4 plasma and the electrode voltage for silicon dioxide samples at different discharge currents: (1) 140 mA, $B = 1$; (2) 120 mA, $B = 0.625$; (3) 80 mA, $B = 0.588$; and (4) 50 mA, $B = 0.555$. Dashed lines indicate relationships (5.35) and (5.40); * indicates software-calculated etch-rate values.

are sufficient only for transporting the ions to the surface and desorbing the CAPs. The zero etch rate at $U = 0.5$ kV indicates that the ion–electron flux at this voltage does not reach the sample surface at all (see Equation 3.8). The cause of the sharp decrease in the etch rate in the voltage range 0.8–1 kV is that the ion energy in this range is sufficient to heat the sample surface: it heats up to a temperature at which the adsorption of CAP-forming CF_4 molecules becomes impossible, thereby resulting in the zero etch rate.

On the whole, the maximum etch rates in the plasma-chemical region are comparable with those given in Reference 274. At voltages of 1.2–2 kV and currents of 120–140 mA, the conditions are ideal for ion-chemical etching because the plasma contains a large concentration of negative ions with an energy on the order of 100–500 eV (see Figure 3.12) that reach the sample: the energy is sufficient not only for ion transport to the surface but also for its sputtering. Along with physical sputtering, chemical interaction is observed in the sputtering region between fluorine ions and sputtered particles, resulting in the formation of gaseous products. This is the reason why during etching there was no deficiency in particles involved in removing the reaction products from the treatment area. This explains why curves 1 and 2 have the largest steepness. And as Figure 5.7 shows, maximum etch rates were observed under these conditions. The subsequent drop in each curve is due to the increasing energy of positive ions bombarding the cathode and to the resulting increase in the secondary-emission coefficient γ_e, leading to an increase in the number of electrons in the plasma. The energies of negative ions and electrons in this voltage range (see Figure 3.12) are high enough within the length between the cathode and sample. The electrons' energy is quite sufficient to neutralize and then ionize the F⁻ ions. In this case the attachment coefficient α_n tends to zero, thereby sharply decreasing the number of negative fluorine ions contributing to etching. Thus, after we transform expression (5.25), the ionization coefficient α can be obtained from the equation

$$\alpha = \frac{1}{d_{max}} \ln\left(\frac{J}{J_e^K}\right) = \frac{1}{d_{max}} \ln\left(\frac{1+\gamma_e}{\gamma_e}\right), \tag{5.43}$$

and J_i^- from Equation 5.28. Note that the electron flux J_e cannot be disregarded because the flux is predominant in this case. In the range 1.2–3.5 kV, curves 3 and 4 are lower than the first two. This is due to a decrease in the number of negative fluorine ions in the plasma as a result of the decreasing discharge current. The zero etch rates in the range 1–1.2 kV indicate that, while the energy of F⁻ ions is not sufficient for sputtering, it is already sufficient for heating the material. Apparently, this region is the transition area of high-voltage gas discharge. For comparison, Figure 5.7 also shows the curves calculated from Equations 5.35 and 5.40 (dashed lines). The close agreement between the calculated and experimental data shows that the models discussed here adequately describe the actual etching process.

Figure 5.8 shows a mask and microrelief profiles on silicon dioxide surfaces (with the mask removed) that were obtained by etching the samples in the plasma-chemical (Figure 5.8b) and ion-chemical (Figure 5.8c) regions.

Figure 5.9a and b shows the surface microreliefs of substrates with the mask removed. These show that, in the plasma-chemical region of the high-voltage gas

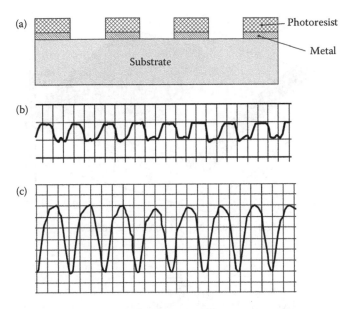

FIGURE 5.8 Mask and microrelief profiles: (a) view of a mask (single-layer photoresist masks and metallized masks were used); (b) microrelief profile of a SiO_2 substrate, obtained in the plasma-chemical region; and (c) microrelief profile of a SiO_2 substrate, obtained in the ion-chemical region. The plasma-forming gas was CF_4. Images (b) and (c) were obtained with the 170311 profilograph/profilometer. One division on the horizontal scale is equivalent to 4 μm, on the vertical scale, to 0.2 μm.

discharge in CF_4 plasma, a microrelief was formed with a period of $T = 12$ μm and a height of 0.2 μm and that the ion-chemical region produced the same period. But the microrelief height formed in that region amounted to 1.2 μm.

This suggests that off-electrode plasma can be used for fabricating optical microreliefs.

To find an optimal plasma-chemical etch rate (V_{pce}), it is important to know how it varies with the oxygen percentage in CF_4/O_2 plasma [190]. For off-electrode CF_4/O_2 plasma, this relationship was studied at different discharge currents (see Figure 5.10). Notice that with an increase in the oxygen percentage in CF_4 plasma, the etch rate first rises and then falls to almost zero values. The curves are similar in shape for all discharge currents except the minimum one, 50 mA.

For this current, the insignificant variation in the etch rate is attributable to the low density of charged particles in the plasma. With a low ionization rate of process-gas molecules by O^- ions, these make a modest contribution to the production of F^- ions; see reactions (5.14), (5.16), and (5.17). With pure CF_4, etching was not observed at the minimum discharge current.

As the discharge current increases, so does the density of charged particles in the plasma.

This, in turn, increases the ionization rate of CF_4 molecules by O^- ions and hence the density of F^- ions produced with the assistance of oxygen. The steep, rising segments of curves 2, 3, and 4 in Figure 5.10 indicate a deficiency in active F^* radicals

(a)

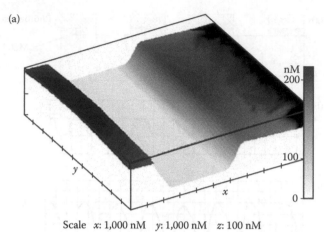

Scale x: 1,000 nM y: 1,000 nM z: 100 nM

(b)

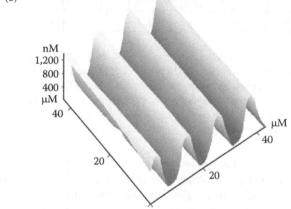

FIGURE 5.9 SiO_2 substrate microrelief: (a) obtained in the plasma-chemical region at $I = 140$ mA and $U = 0.8$ kV and (b) in the ion-chemical region at $I = 140$ mA and $U = 2$ kV. The images were obtained with P4-SPM-MDT (a) and SMENA in combination with NT MDT P47H (b).

at the surface being treated, implying that the etch rate depends on the density of F^- ions. The pronounced peak observed at each discharge current corresponds to the situation in which all of the oxygen takes part in producing F^- ions; at the same time, the oxygen does not compete with F^* radicals for active sites at the surface, nor does it passivate the surface.

Optimal plasma-chemical etch rates are observed at an oxygen percentage as low as 0.5%–1.5% (see Figure 5.10). This finding must indicate high transverse uniformity of the plasma stream, its normal incidence on the surface, and freedom from wall collisions. A targeted delivery of active radicals to the surface takes place: every O^- ion produced in the bulk of the plasma according to Equations 5.10 and 5.12 is involved in the generation of a F^- ion, which, in turn, acts as a CAP. The following reaction chain is observed: $O_2 \xrightarrow{e,F^-} O^- \xrightarrow{CF_4} F^- \xrightarrow{S_n} F^*$.

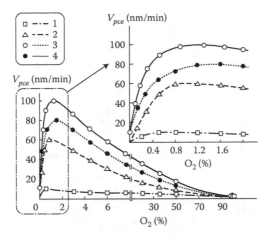

FIGURE 5.10 Relationship between the plasma-chemical etch rate and the oxygen percentage in high-voltage gas-discharge CF_4/O_2 plasma at different discharge currents (at $U = 0.8$ kV) for silicon dioxide: (1) 50 mA, (2) 80 mA, (3) 120 mA, and (4) 140 mA.

An oxygen percentage greater than 1.5% in CF_4/O_2 plasma causes a sharp decrease in V_{pce}. This should be due to the occupation of vacant bonds by oxygen radicals, which thus compete with F* ones. Further, oxygen molecules excited at the surface should react with F* radicals to convert them into F_2, a less reactive substance [275]. The number of CAPs involved in etching is thus reduced.

When the plasma is generated in pure oxygen, the surface is fully passivated so that V_{pce} is close to zero; this conclusion is consistent with established opinion [34,53,95]. Adding oxygen to the main process gas also increases the ion-chemical etch rate in off-electrode plasma. But this behavior is at variance with long-standing views [190,276]. To clarify the point, examine Figures 5.11 and 5.12. On the whole, the ion-chemical etch rate (V_{ich}) follows the same pattern as in the plasma etching case. This is because only neutral process-gas molecules and charged particles are in the bulk of off-electrode plasma—at least along the distance d.

Fluorocarbon and oxygen ions are unlikely to combine into stable molecules (CO, CO_2, and COF_2) on account of the separation of charged particles and the action of a strong, nonuniform electric field (along the distance d) [65].

As a result, high-energy O^- and F^- ions produced in the plasma stream (see reactions (5.10) through (5.14)) should not recombine as they travel toward the surface. These ions will erode the material first by physical sputtering and then by chemical reactions.

In the sputtering, high-energy ions penetrate a certain depth into the material and in doing so break its interatomic bonds. Having lost energy, the ions can interact with the material only through chemical reactions. As with plasma-chemical etching, this stage of reactive ion etching is characterized by competition between reactive fluorine and oxygen species for active sites, with the difference that these are now located in the bulk of the material.

This explains why the etch rate starts falling once the oxygen percentage has exceeded 1.5%–2%. But as Figures 5.11 and 5.12 show, V_{ich} does not vanish however

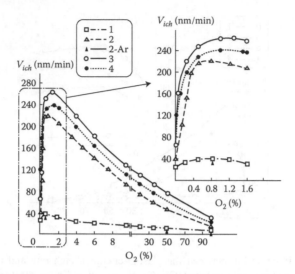

FIGURE 5.11 Relationship between the ion-chemical etch rate and the oxygen percentage in high-voltage gas-discharge CF_4/O_2 plasma at different discharge currents ($U = 2$ kV) for silicon dioxide: (1) 50 mA, (2) 80 mA, (2-Ar) etch rates at a given argon percentage in CF_4/Ar plasma, (3) 120 mA, and (4) 140 mA.

high the oxygen percentage is, implying that pure O_2 etching occurs by physical sputtering with O^- ions. This mechanism starts acting when the O_2 percentage exceeds 10% relative to CF_4.

It is manifested in characteristic dips (called "teeth" in Reference 34) in the etching profile, as shown in Figure 5.13, which indicate that re-evaporation rather than chemical erosion dominates the sputtering [94].

Comparing Figures 5.10 through 5.12, we notice that the etch rate peaks for the same oxygen percentage. But presumably because of the absence of oxygen in the

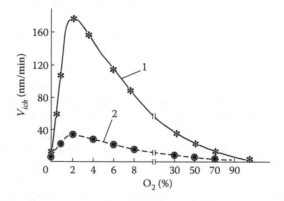

FIGURE 5.12 Relationship between the ion-chemical etch rate ($I = 120$ mA; $U = 2$ kV) of silicon carbide (curve 1) and diamond-like films (curve 2) and the oxygen percentage in CF_4/O_2 plasma generated by high-voltage gas discharge.

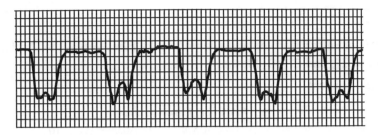

FIGURE 5.13 Ion-chemical etching trench profile at oxygen percentage exceeding 10%. One division on the horizontal scale is equivalent to 2 μm; on the vertical scale, to 0.2 μm.

structures of silicon carbide and diamond-like films, a higher percentage of O_2 (2%) is required for optimal etching than in the case of silicon dioxide. This confirms that in plasma-chemical etching and ion-chemical etching, the same processes occur in the bulk of off-electrode plasma (or at least along the distance d), thus supporting the mechanisms and equations proposed above. Otherwise, adding oxygen would reduce V_{ich} in the case of ion-chemical etching.

The nonzero V_{ich} in pure CF_4, observed even at a discharge current as low as 50–80 mA, signifies that the voltage between the electrodes is the major factor in the transport of active particles to the surface. The higher rate of change shown by the curves in Figure 5.11 is due to physical sputtering. Note that if CF_4 plasma contains argon (see curve 2-Ar in Figure 5.11), which is chemically neutral toward both the material and the main gas, the etch rate reduces to values that are lower than those observed for pure CF_4. This is because an increase in the argon percentage in CF_4/Ar plasma corresponds to an increase in the number of electrons and to a decrease in the number of chemically active fluorine ions moving toward the material. Ar^+ ions move along the force lines toward the cathode and do not enter into the reactions.

Therefore, an oxygen percentage of 0.8%–2% in CF_4 plasma indeed increases the etch rate.

Both in the cases of plasma-chemical etching and of ion-chemical etching, oxygen additions prove the most effective for a discharge current of 80–120 mA (see Figures 5.10, 5.11, and 5.14, curves 1 and 2, and Table 5.1).

A discharge current greater than 120 mA leads to a decrease in the etch rate because of the excessive number of CAPs at the surface impeding the removal of reaction products. In this case the removal of reaction products from the surface becomes a rate-limiting step. Curves 3 and 4 in Figure 5.14, too, show that this is the main mechanism reducing V_{pce} and V_{ich}, and so do the results of experimental investigations into the etching of other materials relevant for diffractive optics (see Table 5.1).

The curves' behavior is exponential and no peak is observed, and this indicates a lack of CAPs at the surface. We can conclude that during plasma-chemical etching and ion-chemical etching in pure-CF_4 off-electrode plasma, the rate-limiting factor is the number of F^- ions generated in the plasma. If the current is increased to a value greater than 140 mA to increase V_{ich}, the resulting overheating of the substrate could destroy the photoresist mask. This was indeed the case as observed in our experiment.

TABLE 5.1

Etch Rate of Silicon Carbide and Diamond-Like Films versus the Discharge Current in CF_4 and CF_4/O_2 Plasma

Discharge Current (mA)	50	80	120	140
Etch Rate in CF_4 Plasma (nm/min)				
Silicon carbide	4.3	6.2	10	18.7
Diamond-like films	4.1	4.4	7.3	13.7
Etch Rate in CF_4/O_2 Plasma (nm/min)				
Silicon carbide	23.1	142.1	175	158
Diamond-like films	4.6	28.4	35	31.6

5.5 RELATIONSHIP BETWEEN THE ETCH RATE AND SUBSTRATE TEMPERATURE

A sample's temperature increases as a result of interaction between plasma and the sample's surface. Substrate temperature can, to a large degree, determine the rates of ion-chemical and plasma-chemical etching: at an optimal temperature under stabilized conditions V_{pce} and V_{ich} can exceed the initial etch rate by a factor of three or more [277]. This results in parameters and characteristics that deviate to a varying degree from those expected for the element being fabricated. Because of this, the relationship between the off-electrode plasma etch rate and substrate temperature is a highly relevant line of research.

To improve the accuracy of micro- and nanostructure fabrication, temperature must be monitored at a site where a plasma flux is incident on the surface, which

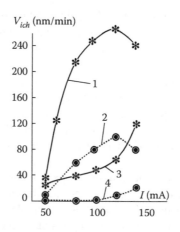

FIGURE 5.14 Relationship between the etch rate of silicon dioxide and the discharge current: (1) ion-chemical etch rate in CF_4/O_2 plasma, (2) plasma-chemical etch rate in CF_4/O_2 plasma, (3) ion-chemical etch rate in CF_4 plasma, and (4) plasma-chemical etch rate in CF_4 plasma.

is difficult given that the electric field of the plasma affects measurements. Optical measurement methods are inapplicable in the high-temperature range and also suffer from nonmonochromatic self-radiation of excited species in gas-discharge plasma.

This section is devoted to developing an analytical method for calculating the surface temperature of a sample bombarded by a directed flux of low-temperature off-electrode plasma and experimentally investigates the relationship between the etch rate and substrate temperature.

5.5.1 METHOD FOR DETERMINING THE TEMPERATURE OF A SURFACE AT A SITE WHERE A LOW-TEMPERATURE PLASMA FLUX IS INCIDENT ON THE SURFACE

Today the numerical methods of calculation find wide application in the theory of heat transfer. Indeed, the problem we are interested in can be viewed as an inverse boundary-value problem of heat conduction. In this case, taking measurements on one part of the surface, one can recover the heat load on other parts inaccessible to measurements. But this inverse problem of mathematical physics belongs to the class of ill-posed problems [278]; therefore, even an approximate solution can be obtained only with special numerical methods [279,280] that provide its stability. At the same time, recent advances in plasma physics make it possible to quantitatively evaluate the effect of plasma in the form of a heat flux on a sample's surface. Therefore, the problem of determining sample temperature is suggested to be reduced to analytically solving the direct problem of heat conduction with mixed boundary conditions.

The charged particles of a plasma flux are uniformly distributed over its cross section in the region where they impinge on the surface. With this in mind and taking into account the geometry of the substrate and its single-crystal structure, we can use the heat-conduction equation for the one-dimensional case [281] with the following boundary and initial conditions [282]:

$$\begin{cases} T(x,0) = T_0 \\ T(b,t) = T_1(t). \\ q(0,t) = q_1 \end{cases} \tag{5.44}$$

As far as we know, analytical solutions to the heat-conduction equation with conditions (5.44) are absent. Therefore, we take the most appropriate known solution to this equation with similar boundary and initial conditions [283]:

$$T(x,t) = \int_0^t q_1(\varepsilon) \frac{\partial \Theta(x,t-\varepsilon)}{\partial t} d\varepsilon + \int_0^t q_2(\varepsilon) \frac{\partial \Theta(b-x,t-\varepsilon)}{\partial t} d\varepsilon + T_0, \tag{5.45}$$

where $q_1(\varepsilon)$ is the specific heat flux incident on the front surface, $x = 0$ (W/m^2), and $q_2(\varepsilon)$ is the heat flux carried away from the back surface of the sample (W/m^2).

The function $\Theta(x,t)$—the temperature response of the body to the unit heat-flux incident on one of the boundaries—is given by [284]

$$\Theta(x,t) = \frac{1}{\lambda} \left\{ \frac{at}{b} + \frac{3(b-x)^2 - b^2}{6b} \right.$$

$$\left. + \frac{2b}{\pi^2} \sum_{k=1}^{\infty} \frac{(-1)^{k+1}}{k^2} \exp\left(-k^2\pi^2 \frac{at}{b^2} \right) \cos\left(k\pi \frac{b-x}{b} \right) \right\}, \qquad (5.46)$$

where b is the sample's thickness (m); $a = \lambda/C$ is the thermal diffusivity (m²/s); λ is the thermal conductivity (W/m·K); and C is the material's heat capacity per unit volume (J/m³·K).

The form of the function $q_2(\varepsilon)$ can be found by measuring the temperature of the sample's lower surface, $T(b, t) = T_{low}(t)$, with the incident heat flux $q_1(\varepsilon)$ known. In this case, the temperature $T(0, t)$ of the exposed surface is a partial solution to the initial Equation 5.45 and depends on $q_1(\varepsilon)$ and the calculated value of $q_2(\varepsilon)$.

Using the Laplace transformation,

$$F(p) = \int_0^{\infty} f(t) \exp(-pt) dt, \qquad (5.47)$$

we represent the left-hand side of Equation 5.45 in the form of the complex variable p,

$$T_{low}(p) \leftarrow T(b,t). \qquad (5.48)$$

The convergence condition imposed on integral (5.47) implies the need for finite approximation of the sum in Equation 5.46. This expression is known to be a convergent alternate series [285]; therefore, the sum can be calculated with the desired accuracy by discarding the right-hand part, and the approximation error is no more than the absolute value of the first of discarded terms.

Let us transform the right-hand side of Equation 5.45 by applying the convolution theorem [286], which allows one to determine the original of the product of images,

$$T_0 + \int_0^t q_1(\varepsilon) \frac{\partial \Theta(x,t-\varepsilon)}{\partial t} d\varepsilon + \int_0^t q_2(\varepsilon) \frac{\partial \Theta(b-x,t-\varepsilon)}{\partial t} d\varepsilon$$

$$\leftarrow T_0(p) + Q_1(p)K_1(p) + Q_2(p)K_2(p), \qquad (5.49)$$

where $T_0(p)$, $Q_1(p)$, and $Q_2(p)$ are the images of initial temperature T_0 and heat fluxes $q_1(\varepsilon)$ and $q_2(\varepsilon)$, respectively, and $K_1(p)$ and $K_2(p)$ are the images of the time derivatives of temperature responses $\Theta(b,t)$ and $\Theta(0,t)$, respectively,

$$\begin{cases} K_1(b,t) = \dfrac{a}{b\lambda}\left[1+2\sum_{k=1}^{n}(-1)^k \exp\left(-\dfrac{k^2\pi^2 at}{b^2}\right)\right], \\[4mm] K_2(0,t) = \dfrac{a}{b\lambda}\left[1+2\sum_{k=1}^{n}(-1)^{k+1} \exp\left(-\dfrac{k^2\pi^2 at}{b^2}\right)\right]. \end{cases} \qquad (5.50)$$

With regard to Equations 5.48 and 5.49, the desired solution for $Q_2(p)$ takes the form

$$Q_2(p) = \frac{T_{low}(p) - T_0(p) - Q_1(p)K_1(p)}{K_2(p)}. \qquad (5.51)$$

To find the surface temperature, we will substitute the known value of $q_1(\varepsilon)$ and original $Q_2(p)$, calculated with Equation 5.51, into the initial Equation 5.45. Note that, in going from function $T(b, t)$ to the desired function $T_{surf}(t) = T(0, t)$ (i.e., surface temperature), the forms of $\theta(x, t)$ and $\theta(b-x, t)$ change since the coordinate $x = b$ is replaced by $x = 0$. With this in mind, we can write Equation 5.45 in the complex form

$$T_{surf}(p) = T_0(p) + Q_1(p)K_2(p) + Q_2(p)K_1(p). \qquad (5.52)$$

The temperature of the exposed surface can be found from Equations 5.51 and 5.52:

$$\begin{aligned} T_{surf}(p) &= T_0(p) + Q_1(p)K_2(p) \\ &+ \frac{K_1(p)}{K_2(p)}(T_{low}(p) - T_0(p) - Q_1(p)K_1(p)). \end{aligned} \qquad (5.53)$$

Expression (5.53) shows that if the sample is thin ($b \to 0$), T_{surf} approaches T_b. Indeed, as follows from Equation 5.50, the expression under the summation sign is an infinitesimal; in this case, $K_1 = K_2$. Substituting this equality into Equation 5.53 gives a negligibly small temperature gradient in a plane sample.

The real function $T_{surf}(t)$ is found with the inverse Laplace transformation [286]

$$T_{surf}(t) = \int_{-\infty}^{+\infty} T_{surf}(p)\exp(pt)\mathrm{d}p. \qquad (5.54)$$

Thus, using the integral transformations, we have derived the expression for the sample temperature in the region exposed to a directed flux of low-temperature plasma as a function of the known parameters. The disadvantage of our method is the difficulty of going to Equation 5.53. But this problem can be completely eliminated with mathematical program packages.

The model proposed was used to calculate the surface temperature of a silicon dioxide substrate exposed to off-electrode plasma irradiation. In the near-surface

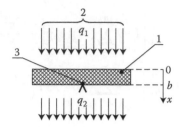

FIGURE 5.15 Irradiation of the sample by the gas-discharge plasma flux: (1) dielectric substrate, (2) directed plasma flux, and (3) temperature sensor at the lower surface.

layer, the plasma flux incident on the substrate surface produced heat flux q_1, which, having passed through the substrate, turned into flux q_2 (Figure 5.15). The lower-surface temperature was measured with a precision chromel–copel thermocouple.

The lower-surface temperature was found not to exceed 700 K (its variation is presented in Figure 5.16).

The temperature curve was interpolated by a polynomial whose order was determined to a given accuracy. In this temperature range, the mean values of the thermophysical parameters of the substrate were taken to be $a = 0.00001$ m^2/s and $\lambda = 10$ W/(m·K) [287].

The value of q_1 is calculated with the techniques presented in Reference 209 as the product of the cathode-emitted electron flux and the electron energy determined by the accelerating voltage. The assumption is valid since the ion component of the plasma has a low energy as compared to the electron one, and the ions rapidly lose energy in collisions with process-gas atoms. On the other hand, at a pressure of 10^0–10^{-1} Torr and a cathode–substrate distance of $d = 0.05$ m, the number of elastic collisions of electrons with process-gas atoms is small and the energy loss is insignificant.

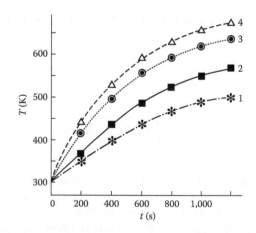

FIGURE 5.16 Temperature change at the lower surface: (1) $I = 50$ mA, (2) $I = 80$ mA, (3) $I = 120$ mA, and (4) $I = 140$ mA. Electrode voltage $U = 2$ kV. Pressure $p = 1.5 \cdot 10^{-1}$ Torr. The process gas is halocarbon-14 (CF_4).

It is known that whether series (5.46) converges depends on the value of at/b^2: the greater this parameter, the better the convergence. To find an exact solution at small at/b^2 (e.g., at the initial stage of the process), it is necessary to use 11–12 terms of the series [288]. In this study, we took into account 12 terms of sum (5.46).

As was noted earlier, the boundary-value problem is rather difficult to solve analytically because Equation 5.53 contains the ratio of series K_1 and K_2. Because of this, the proposed algorithm was implemented by using the Maple 8 software package. With Equation 5.53, we constructed graphs that show the relationship between the temperature gradient ΔT in the substrate and the process time (Figure 5.17).

As Figure 5.17a shows, the curves first ascend sharply. This is because the substrate, being thin, heats up rapidly. The incident flux $q_1(\varepsilon)$ passes through the sample almost instantly without noticeable energy loss and leaves the lower surface, rapidly causing a temperature difference. When the irradiation time is long, the sample heats up at a constant temperature gradient (Figure 5.17a and b).

It is this circumstance that may be responsible for the thermal shock effect [289] when thin samples are almost instantly destroyed once the discharge power exceeds a critical value. Indeed, thermal stresses that occur are determined by the temperature gradient, which rapidly runs through its intermediate values and reaches a maximum virtually at the very beginning of the process (see Figure 5.17a). The simulation data suggest that the transient time increases as the thermal diffusivity of the sample decreases or it gets thicker, with the thermal action $q_1(\varepsilon)$ being the same. It is evident that this statement completely agrees with the theory of heat transfer: a more massive sample reaches the stationary state for a longer time. And a material with a lower thermal conductivity will have a higher temperature gradient, which will be established for a longer time. The rigorous solution to this problem implies a combined consideration of the equations of heat transfer and thermoelasticity [281].

At high t, the temperature difference takes on a constant value (Figure 5.17b). Therefore, the sample is unlikely to fail at the final stages. The model proposed was also experimentally verified on KEF-32 silicon samples measuring $0.01 \times 0.01 \times 0.001$ m³. The temperature of the samples was controlled by varying

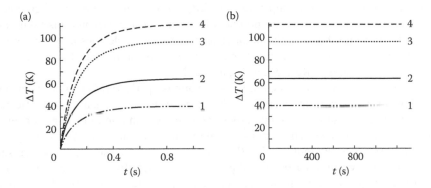

FIGURE 5.17 Temperature difference between the upper and lower surfaces for a bombardment time of (a) 1 s and (b) 1,200 s. (1) $I = 50$ mA, (2) $I = 80$ mA, (3) $I = 120$ mA, and (4) $I = 140$ mA.

TABLE 5.2

Exposed-Surface Temperatures and Temperature Gradients

Temperature of the exposed surface (K)	1,063	1,173	1,183	1,198
Experimentally obtained gradient (K)	15	10	10	10
Theoretically obtained gradient (K)	16.5	10.8	11.1	11.2

the parameters of plasma-flux irradiation: voltage from 2.6 to 5.3 kV and current from 24 to 80 mA. The irradiation time was 10 min. The thermo-physical parameters of the material were matched to the process conditions. The temperatures of the exposed surface were measured with a Promin micropyrometer. The surface temperatures and temperature gradients are listed in Table 5.2.

The discrepancy between the calculated and experimental values of the temperature difference does not exceed 12%, and this confirms the adequacy of the estimation method. The somewhat higher theoretical values might indicate that the solid does not completely absorb the plasma flux: part of the flux is reflected from the exposed surface, thereby decreasing the gradient.

The proposed method makes it possible to monitor the surface temperature during etching in a directed low-temperature plasma flux. Using the method will help improve the fabrication quality of micro- and nanostructures by stabilizing thermally unstable fabrication processes and optimizing the low-temperature plasma etch rate.

5.5.2 EXPERIMENTAL INVESTIGATION INTO THE RELATIONSHIP BETWEEN THE ETCH RATE AND SUBSTRATE TEMPERATURE

The method discussed in Section 5.5.1 [290] was used to monitor the surface temperature of samples in an experimental investigation into the relationship between the etch rate and substrate temperature in off-electrode plasma. For plasma-chemical etching, the relationship yielded the curves shown in Figure 5.18.

The experimental results show that for all discharge currents, the etch rates peak at 360 K, the vaporization temperature of SiF_4. At this temperature, perfect conditions arise for removing reaction products from the surface. Beyond this point, the curves begin to fall because the number of process-gas molecules adsorbed at the surface diminishes with an increase in substrate temperature, and this agrees well with the observations in References 212 and 252.

In the case of ion-chemical etching, the relationship $V_{ich} = f(T)$ is more complex. At a low discharge current (curve 1 in Figure 5.19), $V_{ich} = f(T)$ is almost unaffected by the variation of substrate temperature because the etch rate in this case depends on the density of F$^-$ ions. This agrees well with the conclusion drawn earlier.

At $I = 50$ mA in the range $325 < T < 360$ K, etching is possible because the surface is almost free from particles that could impede etch-product removal from the surface layer by evacuation facilities. At a discharge current of $I > 50$ mA (curves 2, 3, and 4 in Figure 5.19), the reverse is observed: the particles (F$^-$ ions, CAPs, and interaction products) accumulated at the surface and in the near-surface layer impede the removal of reaction products (mostly SiF_4). As a result, etching becomes possible

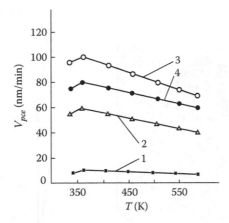

FIGURE 5.18 Relationship between the plasma-chemical etch rate and SiO_2 substrate temperature in high-voltage gas-discharge CF_4/O_2 plasma for different discharge currents: (1) 50 mA, (2) 80 mA, (3) 120 mA, and (4) 140 mA.

only at temperatures that are equal to or exceed the volatilization temperature of silicon tetrafluoride.

This explains the similar progress of the curves in the range $300 < T < 360$ K. With a further increase in temperature, V_{ich} increases, and for any given discharge current there is a corresponding temperature at which the etch rate reaches its maximum value. This is because an increase in temperate weakens interatomic bonding: high-energy ions penetrate to a greater depth and thus cause more atoms to sputter off the material.

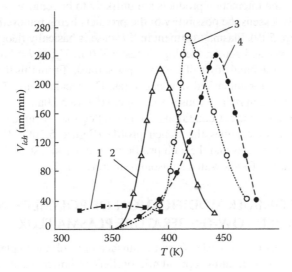

FIGURE 5.19 Relationship between the ion-chemical etch rate and SiO_2 substrate temperature in CF_4/O_2 plasma for different discharge currents: (1) 50 mA, (2) 80 mA, (3) 120 mA, and (4) 140 mA.

(a)   (b)

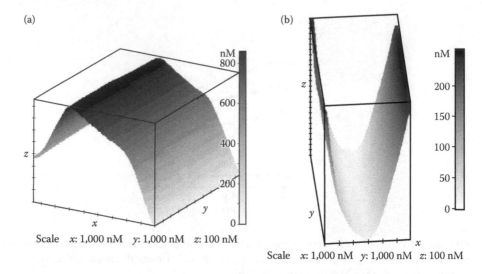

Scale *x*: 1,000 nM *y*: 1,000 nM *z*: 100 nM

Scale   *x*: 1,000 nM *y*: 1,000 nM *z*: 100 nM

FIGURE 5.20 Etch taper formed by photoresist-mask breakdown (a) and the associated trench profile (b). $I = 140$ mA; $U = 2$ kV; substrate temperature = 440 K.

On the other hand, the higher the discharge current, the more ions penetrate the sample's near-surface layer to form reaction products that impede one another's reaching the surface. Higher temperatures are therefore required to remove the products. The sharp fall in the ion-chemical etch rates at temperatures greater than $T > 390$ K, $T > 422$ K, and $T > 440$ K (corresponding to discharge currents of $I = 80$ mA, $I = 120$ mA, and $I = 140$ mA, respectively) apparently results from ions penetrating to a depth at which the interaction products are unlikely to be removed—an increase in temperature only lessens the possibility of the products being removed (see curves 2, 3, and 4 in Figure 5.19). Plasma treatment in this case is basically fluorine-ion doping of the surface layer and sputter etching. A faster fall in V_{ich} at $I = 140$ mA is, in all likelihood, due to the breakdown of the photoresist mask. From this time onward, the etch rate is the same for unmasked and open areas. The breakdown at $T = 440$ K starts from the edges of the mask and causes etch taper (Figure 5.20a).

The taper creates perfect conditions for ion reflection—it guides ions just into trenches and thus determines the trench profile (Figure 5.20b). The etch taper's growth smooths its angles and the etch profile becomes a sinusoid. This property is useful for fabricating DOEs with a sinusoidal microrelief [1].

5.6 EFFECT OF BULK MODIFICATION OF POLYMERS IN A DIRECTED LOW-TEMPERATURE PLASMA FLUX

Removing polymers (resist masks) from the surface of a solid with plasma etching is the final operation in fabricating optical microreliefs and micro- and nanostructures. The regularities of this process have been studied for a long time [11,108,236,250,251]. But despite the large number and apparent comprehensiveness of the available experimental results, the mechanism of polymer etching is not completely clear because of its complex multifactor dependence on the type of interaction between active particles in the plasma and the polymer matrix.

In the plasma-chemical etchers that have been used thus far, plasma is generated by gas discharge in the electrode gap—see, for example [35,94]. For purposes of this monograph, we have created a reactor in which high-voltage gas discharge generates low-temperature plasma outside the electrode gap. Similar plasma generators have proved promising for welding [63], soldering semiconductor components [64], cleaning material surfaces [66], and enhancing thin-film adhesion [291,292].

This section is devoted to our experimental investigation into the regularities of polymer etching in the plasma generated outside the electrode gap in oxygen. The experimental results we obtained provided the basis for creating a computational model of the etching process.

Figure 5.21 shows the experimental relationship between the thickness of the scoured polymer layer (h) and the etching time (t) for two different values of the initial film thickness. Both curves display identical behavior in the region $0 \leq t \leq 18$ s: h increases for $0 \leq t \leq 6$ s and $15 \leq t \leq 18$ s ($15 \leq t \leq 21$ s for curve 1) and the etch rate decreases for $6 \leq t \leq 15$ s. Both curves have saturation regions for h values equal to the corresponding film thicknesses, and this confirms the complete removal of polymer from the surface.

We will use the experimental results for creating a model for polymer etching in the oxygen plasma outside the electrode gap.

The most comprehensive mechanisms and models for polymer etching in HF and SHF plasma have been proposed in References 11, 250, and 251. These studies assume that a modified surface layer—a K-layer—is formed during etching and that it is more resistant to destruction than unmodified lower layers of the polymer structure.

But the K-layer model was developed on the basis of experiments on etching in electrode plasma. Interpreting our results on etching outside the electrode gap allows us to supplement this model by the idea that the modified layer in this case may lie in the bulk of the polymer.

In oxygen plasma, atomic oxygen (Ö), negative oxygen ions (O⁻), and excited molecular oxygen (O_2^*) with a low concentration on the order of $10^{-2}\%$ are active

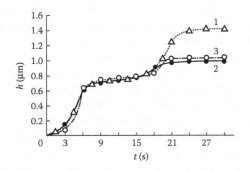

FIGURE 5.21 Relationship between the thickness of the scoured polymer layer and the etching time for $I = 100$ mA and $U = 2$ kV: (1) initial thickness of the polymer film is 1.4 μm and (2) 1 μm; (3) calculated relationship for an initial thickness of 1 μm.

etching particles [190]. Polymer etching may occur through sputtering by high-energy O^- ions, as well as through their chemical interaction with polymer molecules. The atomic oxygen Ö present at the surface can also interact with these molecules. The reaction products form the volatile compounds H_2O (water vapor), CO_2, and N_xO_y, which are removed from the working chamber by evacuation facilities.

The role of electrons in this process is controlled by the following circumstance: An electron's mean free path in the gas and in the polymer is much greater than the mean free path of an ion, because of the fewer collisions with atoms and molecules of the medium. Electrons penetrate to the bulk of the polymer to the depth [87]

$$L = 10^{-5} \frac{U^{3/2}}{\rho}, \tag{5.55}$$

where $\rho = 0.5$ g/cm^3 is the polymer density; $U = 2$ kV is the accelerating voltage; and $L = 0.57$ μm, which is half the thickness of the polymer film and agrees with experimental curve 2 (see Figure 5.21). Electrons are decelerated in the substance because of the excitation of atoms in the polymer molecules. In each collision, an electron expends for excitation the energy [94]

$$\varepsilon = \frac{2m_e}{M} E_e, \tag{5.56}$$

where M is the mass of an atom in a polymer molecule and E_e is the electron's initial energy. For $E_e = 2,000$ eV, $\varepsilon \approx 0.005$ eV, which is several orders of magnitude lower than the ionization loss. The distribution of the electron's energy loss over the path depth in this case can be described by the Thomson–Widdington law [293]. An electron experiences about 30 collisions over the length L; in this case, it releases an energy of 1.9 keV at the end of its path, spending this energy to rupture the bonds between the atoms in the polymer layer.

As a result of excitation, polymers may experience relaxation, which is observed at temperatures equal to or exceeding the glass-transition temperature T_s [294]. For a diazonaphthoquinone/novolac (DNQ) resist obtained from metacresol novolac, $T_s = 423$ K [108]; consequently, relaxation does not take place. Hence, the increase in the dependences in the region $0 \le t \le 6$ s can be explained by the interaction between active plasma particles and excited polymer atoms, for which the number of active bonds, N_a, is determined by the flux of electrons, their energy E_e, and time t of the process.

When the rupture of atomic bonds takes place, atoms containing a single uncompensated electron each on the outer orbital try to fill the orbital. Bonds involving the collectivization of electron pairs are formed between adjacent carbon atoms. Thus, a modified layer consisting predominantly of carbon atoms is formed at a depth of L. This layer must possess an elevated density, ρ_m, as compared with unmodified layers, and resistance to destruction [250,251]. How homogeneous this layer is depends on how uniformly the charged particles are distributed over the cross section of the plasma flux and on the dose and energy of electron irradiation calculated for the number of carbon atoms in the layer with different numbers of ruptured (suppressed) bonds and, accordingly, with different degrees of modification (Figure 5.22a).

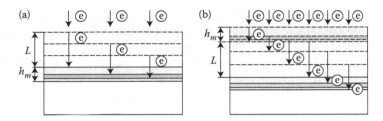

FIGURE 5.22 Schematic illustrating the formation of a modified layer by electrons: (a) etching of the polymer with initial properties and (b) etching of the modified polymer layer.

This mechanism explains the first two regions for $0 < t < 6$ s and $6 < t < 15$ s of curve 1 in Figure 5.21. For $15 \leq t \leq 21$ s, curve 1 in Figure 5.21 has a second segment in the dependence of $h = f(t)$, indicating the etching of a material with properties close to initial properties. Let us consider the mechanism of its formation.

The motion of electrons in a denser medium is accompanied by their scattering, which is proportional to the mean free path. As the modified polymer layer is etched, the mean free path decreases, and this increases the electron flux and energy (ΔE_e) carried by the electron flux to the lower (unmodified) region.

This becomes possible if the etch rate V_m in the modified layer exceeds its formation rate V. In this case, if the condition $\Delta E_e \geq E_{thr}$ is satisfied (where E_{thr} is the threshold energy of delocalization, which is a part of the binding energy [295]), a new stage of layer formation with different degrees of modification begins (it includes the stage of atom excitation) (Figure 5.22b). The number of such layers is proportional to the thickness of the polymer film. The correctness of these statements follows from experimental curve 1. Indeed, this curve clearly displays the second peak corresponding to the stage of forming the second modified layer.

Thus, polymer removal comprises two stages: etching of unmodified and modified layers. The second stage for an individual region of the polymer lags behind the first stage by t_m, the etching time for the unmodified polymer.

Let us estimate the height h of the etched layer as a function of parameters of the physical process (discharge current, accelerating voltage, and etching time) on the basis of the proposed mechanism and experimental results. The value of h is [296]

$$h = \sum_{n=0}^{l-1} \left[\int_{nT}^{t_m+nT} V_0(t)\,dt + \int_{t_m+nT}^{(n+1)T} V_m(t)\,dt \right], \tag{5.57}$$

where $T = t_m + t_k$ (t_k is the etching time of the modified polymer); $n = 0, 1, 2, \ldots, l-1$ (l is the number of modified layers); and t is the etching time. Given that excitation of polymer atoms increases the etch rate and that the decrease in this rate is due to rupture (suppression) of bonds in the polymer at a depth of L, we can write

$$V_0(t) = V_0 \frac{N_a(t)}{N_{sn}}; V_m(t) = V_m \left(1 - \frac{N(t)}{N_{sm}} \right), \tag{5.58}$$

where N_{sn} and N_{sm} are the total numbers of bonds in unmodified and modified layers with thicknesses equal to L and h_m, respectively; N is the number of ruptured (suppressed) bonds; and V_0 and V_m are the etch rates for the polymer and the modified layer, respectively, in high-voltage gas-discharge plasma, and they equal [209]

$$V_0 = \frac{BM}{\rho N_A} J_i^- \left| \exp\left(\frac{U - U_{gd}}{U} \right) - 1 \right| (k_1 + k_3),$$

(5.59)

$$V_m = \frac{BM}{\rho_m N_A} J_i^- \left| \exp\left(\frac{U - U_{gd}}{U} \right) - 1 \right| (k_1^m + k_3^m),$$

(5.60)

where k_1 and k_1^m are plasma-chemical etching coefficients equal to the number of polymer atoms in the unmodified and modified layers removed by one CAP; and k_3 and k_3^m are physical-sputtering coefficients equal to the number of atoms knocked out from the surface of the unmodified and modified layers by one bombarding particle.

Having analyzed the manner in which the experimental curves change in the region $0 \le t \le 6$ s (see Figure 5.21) and the statement concerning the dependence of N_a on the electron flux and energy, as well as on the duration of the process, we approximate $N_a = f(J_e, E_e, t)$ with an exponential function of the form

$$N_a(t) = N_0 \exp\left(\frac{J_e S E_e}{E^*} t \right),$$

(5.61)

where N_0 is the number of bonds at the polymer surface (on the order of 10^{16}); $E^* = N_{sn} E_{thr}^*$ is the total energy required for exciting polymer atoms in a layer with a thickness of L; E_{thr}^* is the excitation threshold energy of a polymer atom; J_e is the electron flux; S is the area of interaction between the low-temperature plasma and the polymer; and E_e is the electron energy.

Analysis of the DNQ-resist structure based on diazoquinone and metacresol novolac (see Figure 5.23) leads to the conclusion that carbon is the main bond-forming element. Knowing the number of carbon atoms n_C in a polymer molecule

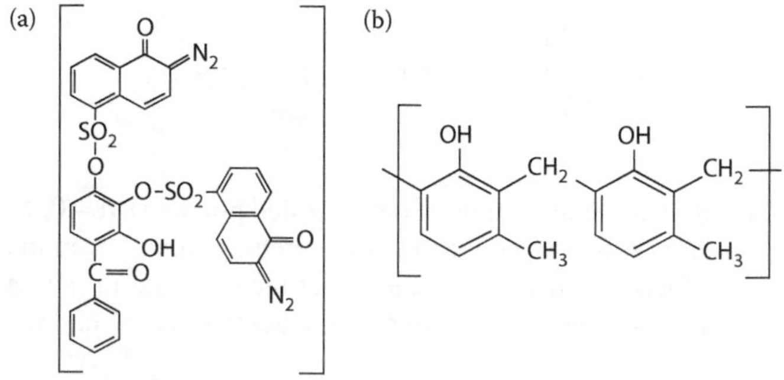

FIGURE 5.23 Molecular structure of DNQ resist: (a) molecular structure of diazonaphthoquinone and (b) molecular structure of novolac.

and carbon valence V_C, as well as the total number of atoms n_{at} in a polymer molecule, we can calculate N_{sn} and N_{sm} from the equations

$$N_{sn} = (V_C n_C) \frac{\rho S}{M} \frac{N_A}{n_{at}} L,$$

$$N_{sm} = \frac{N_{sn}}{L} h_m; \quad h_m = 10^{-5} \frac{U^{3/2}}{\rho_m}. \tag{5.62}$$

Diazoquinone   Novolac resin

Substituting the known values of $n_C = 47$, $V_C = 4$, and $n_{at} = 108$ into these equations, we obtain $N_{sn} \approx 0.4 \cdot 10^{18}$ and $N_{sm} \approx 0.16 \cdot 10^{18}$. The energy released by an electron at the end of its path in the polymer for rupturing (suppressing) the bonds is controlled by the difference $E_e - E^*$, where E^* is the total energy spent by the electron to excite the polymer atoms. In this case, N can be represented similarly to Equation 5.61 in the form

$$N(t) = N_0 \exp\left(\frac{J_e S}{N_{sm}} \frac{(E_e - E^*)}{E_{thr}} t \right). \tag{5.63}$$

In the time interval $0 \le t \le t_m$, etching of the polymer with initial properties takes place; as a result, k_1^m and $k_3^m = 0$ and the second term in Equation 5.57 vanishes. The thickness of the scoured layer is proportional to the number of active bonds of excited polymer atoms. The ratio N_a/N_{sn} specifies the law of variation of the value of h in the segment $0 \le t \le 6$ s in the relationship $h = f(t)$ depicted in Figure 5.21. But modified layers with various degrees of modification are formed at a depth of $h \ge L$. By the instant $t = t_m$, for which the number of active bonds becomes equal to the number of bonds in the unmodified layer ($N_a = N_{sn}$), etching of the unmodified polymer is completed, and this leads to zero values for k_1 and k_3. For $6 \le t \le 15$ s, polymer layers with various degrees of modification experience etching; an increase in N slows down this process (this does not contradict the above mechanism). The law of variation of h at this segment of the curve $h = f(t)$ specifies the ratio N/N_{sm} subtracted from unity.

The instant corresponding to the completion of the modified layer's etching is governed by the equality $N = N_{sm}$, which leads to the vanishing of the second term in Equation 5.57. Alternation of the conditions for completing the etching of the modified and unmodified layers in time at the instant when these conditions hold makes it possible to use expression (5.57) for estimating the value of h for an arbitrary thickness of the polymer film for the given values of the discharge current, accelerating voltage, and t.

To obtain numerical values of N_a and N and, therefore, the thickness of the scoured layer after substituting expressions (5.58) into (5.57), we must know the threshold values of excitation energy and delocalization energy, as well the total energy spent by an electron for exciting the polymer atoms. The results of computer and natural experiments lead to the conclusion that the calculated relationship approximates experimental curve 2 (see Figure 5.21) if $E_{thr}^* \approx 0.005\,\text{eV}$ and $E_{thr} \approx 0.015$ eV; in this case, the value of E^* varies from hundreds of electron volts to dozens to zero, which is due to the decrease in the electron's mean free path in an unmodified polymer during its etching. These values of E_{thr}^* and E_{thr} are two or three orders of magnitude lower than the ionization energy and satisfy the inequality $E_{thr}^* < E_{thr} < E_b$ (where E_b is the binding energy), and this does not contradict the physical process and established opinion [295].

The experiments confirm that a modified layer can exist not only in the near-surface area of the polymer, as reported in References 11, 250, and 251, but also in its bulk at the depth to which electrons penetrate the material.

Thus, with the Thomson–Widdington law in mind, we have proposed and substantiated a mechanism that proceeds from a single perspective to explain the kinetics of polymer etching and formation of modified layers in plasma generated by a high-voltage gas discharge outside the electrode gap. We have discovered polymer bulk modification, an effect that broadens the understanding of processes accompanying interaction between low-temperature plasma and the polymer. The mechanism we have discussed can be used for describing the etching of other materials with low-temperature plasma as well as multicomponent plasma.

5.7   ETCHING QUALITY OF OPTICAL MATERIALS

For each given discharge current at oxygen concentrations corresponding to maximum etch rates, microreliefs were obtained with periods of $T = 12$ μm (Figure 5.24a, b, and e) and 32 μm (Figure 5.24c, d, and f) and heights of $h_m = 0.2$ μm (Figure 5.24a and c) and 1.1 μm (Figure 5.24d and f).

Key quality criteria comprised the deviation of step sidewalls from the vertical, etching nonuniformity across the substrate surface, and the surface roughness of the fabricated microrelief. Before photoresist stripping, the etched samples were examined and found to be free from etch undercut (masked areas did not change in size before and after etching). This indicated the anisotropy of off-electrode plasma etching. Similar examinations were carried out with the Philips XL 40 microscope. In scanning electron microscopy, the incidence angle of a primary-ion beam is the key parameter determining the yield of secondary electrons. With this in mind, we used samples with photoresist masks having step sidewalls inclined at an angle of about 18° (see Figure 5.25a).

Figure 5.25b shows the microrelief of a silicon carbide substrate that was measured with a scanning electron microscope. The microrelief's profile geometry recreates that of the mask, suggesting that the mask's surface pattern has been transferred onto the substrate surface and thus indicating anisotropic etching. As Figure 5.24 shows, the profile approaches a vertical-walled pattern with growing discharge current, as predicted earlier.

For instance, off-electrode plasma with a current of 50 mA is deficient in F⁻ ions, but these rarely collide with process-gas molecules and so have energies as high as 100–500 eV (see Equation 3.7). Favorable conditions thus arise for the reflection of F⁻ ions from trench sidewalls toward the center of the bottom, an area where most of

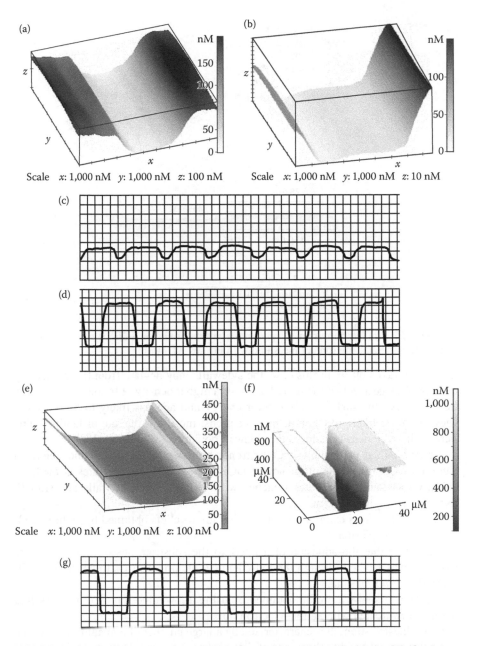

FIGURE 5.24 Images of trenches obtained by etching silicon dioxide in CF_4/O_2 plasma at different discharge currents and optimal oxygen percentage. For (a) and (c), $I = 50$ mA and $O_2 = 0.5\%$; one division on the horizontal scale of the profilogram is equivalent to 2 μm and on the vertical scale, to 0.2 μm. For (b) and (d), $I = 80$ mA and $O_2 = 0.8\%$; and for (e), (f), and (g), $I = 120$ mA and $O_2 = 1.3\%$; one division on the horizontal scale is equivalent to 4 μm and on the vertical scale, to 0.2 μm. $U = 2$ kV for all cases.

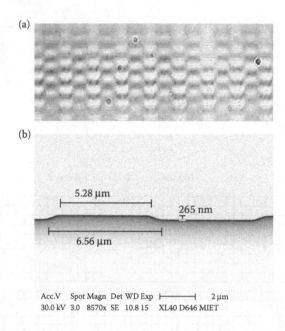

FIGURE 5.25   (a) Interferogram of a photoresist mask formed on a silicon carbide substrate (a) and the associated microrelief measured with Philips XL 40 (b).

the etching takes place. In this case, the sidewalls may deviate from the vertical by an angle as large as 80–70° (Figure 5.24a). At a higher density of F^- ions ($I = 80$ mA), the ions strike the surface with a lower energy and are less likely to be reflected. They are more likely to enter surface reactions, mostly at the site of landing. The increase in CAP density reduces the sidewall deviation [236] to 60–30°.

Figure 5.24b, e, and f shows that the trench bottoms, being smooth and free from acute angles, improve the efficiency of DOEs. Etching SiO_2 substrates at $I = 120–140$ mA was found to produce trenches with virtually vertical walls and smooth bottoms (see Figure 5.24f and g).

Figure 5.26 shows examples of diffraction microreliefs obtained by etching SiO_2 in off-electrode plasma.

The effect that the adhesion of a mask to the substrate has on microstructure geometry was studied through the ion-chemical etching of SiO_2 trenches with off-electrode CF_4/O_2 plasma.

Figure 5.27 shows a profile obtained by using a low-adhesion chrome mask that had not been treated with an ion–electron flux.

Low adhesion strength is characterized by a large number of a metal film's surface atoms that do not bond to the atoms of the substrate material. During etching in gas-discharge plasma, the accumulation of energy intensifies in such areas, thereby causing the mask to break down. The breakdown begins at the mask's edges simultaneously along the entire metal–substrate interface, as suggested by the smooth step angles, and causes a significant etch undercut in the substrate material and irreversible changes in the trench shape. Metal particles sputtered by negative fluorine ions are adsorbed at the trench bottom and cause tapered defects, thereby impeding uniform etching.

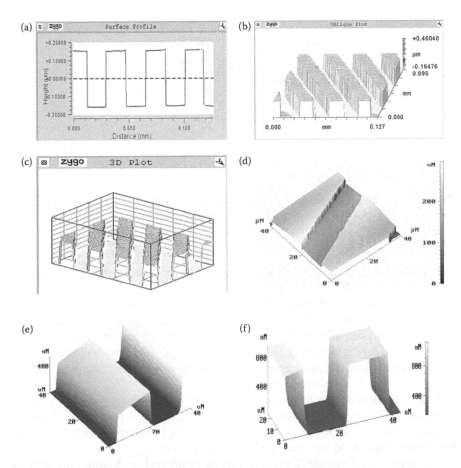

FIGURE 5.26 Examples of diffraction microreliefs obtained by etching SiO_2 in CF_4/O_2 plasma: the profile in (a), (b), and (c) were measured with the Zygo NewView-5000 micro-interferometer; the same profile in (d) and the profiles in (e) and (f) were measured with the SMENA scanning probe microscope combined with NT MDT P47H. Etching conditions: (a–d) $I = 120$ mA, $U = 2$ kV, $O_2 = 1.3\%$; (e) and (f) $I = 80$ mA, $U = 2$ kV, and $O_2 = 0.8\%$.

The desired geometry of microstructures was achieved through the use of masks pretreated with an ion–electron flux in argon. This resulted in trenches with almost vertical walls and smooth bottoms without any irregularities (see Figures 5.24f and 5.26e and f). The masks maintained resistance to plasma throughout the etching despite the discharge current as high as 120 mA (Figure 5.24f). On average, after they were bombarded by an ion–electron flux, the masks remained plasma-resistant for 40–60 min at a discharge current of 80 mA and an accelerating voltage of 2 kV.

Microrelief parameters were measured with both Zygo NewView-5000 and SMENA combined with NT MDT P47H. Since this high-tech equipment is highly precise, the measurement results can be accepted as reliable.

The above regimes were tested for etching two other materials used in diffractive optics and microelectronics—silicon carbide and diamond-like films. Table 5.3

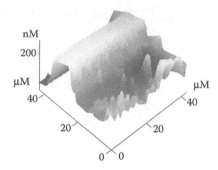

FIGURE 5.27 Substrate microrelief obtained through ion-chemical etching with high-voltage gas-discharge CF_4/O_2 plasma by using an untreated chrome mask. Etching conditions: $I = 80$ mA, $U = 2$ kV, and t (etching time) = 70 s.

lists maximum etch rates in off-electrode CF_4/O_2 plasma for these materials at $I = 120$ mA, $U = 2$ kV, and $O_2 = 2\%$.

These values are comparable with those in References 35 and 53. With this in mind, we can recommend using off-electrode plasma for anisotropic etching of trenches in diffractive optics and microelectronics. The quality of the trenches shown in Figures 5.24, 5.26, and 5.28 is high enough for the techniques of off-electrode plasma etching to be recommended for opening the mask windows of integrated circuits as well as for fabricating diffraction microreliefs.

A fairly thick deposit is formed on the cathode during etching (see Figure 5.29).

Figure 5.30 is an x-ray diffraction pattern [297] from the deposit; it indicates that the deposit contains elements from the process gas (C), the etched materials (SiO_2, SiC, Si, and C), and the etch masks (Cr_2O_3, CrO_3, C, and H_2).

The deposit also includes large amounts of compounds containing the cathode material and different oxides. On the other hand, it is free from fluorine, a fact suggesting that fluorine is totally involved in etching (as part of CAPs), and the presence in the deposit of compounds such as SiO_2, SiC, Si, and C implies that the plasma ensures etch-product removal. It follows that the working plasma species (F^- ions) move toward the substrate and the product ones, toward the cathode. This result supports the mechanisms presented above and agrees with earlier research [209].

TABLE 5.3
Etch Rates for SiO_2, SiC, and Diamond-Like Films in High-Voltage Gas-Discharge CF_4/O_2 Plasma

Material	Etch Rate (nm/min)
Silicon dioxide (SiO_2)	266
Silicon carbide (SiC)	175
Diamond-like films	35

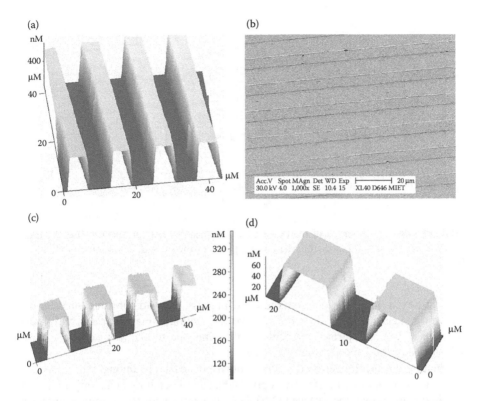

FIGURE 5.28 Microrelief obtained with high-voltage gas-discharge CF_4/O_2 plasma: (a)–(b) SiO_2 surface, (c) SiC surface, and (d) diamond-like film. Measured with the SMENA scanning probe microscope (combined with NT MDT P47H) and Philips XL 40.

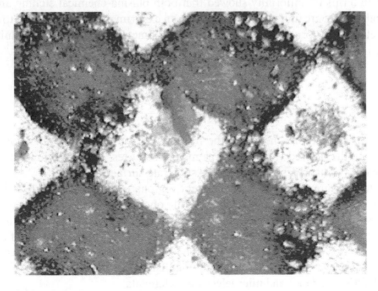

FIGURE 5.29 Cathode surface after etching, ×36.

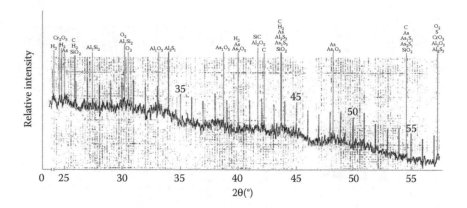

FIGURE 5.30 Radiograph of the deposit adsorbed at the cathode surface; θ is the angle at which x-rays are reflected from the atomic plane of the crystal. The x-ray wavelength is 0.154 nm.

Thus, even with a highly contaminated process gas and substrate surface, off-electrode plasma etching does not involve interactions other than a useful one— between CAPs and surface molecules—allowing one to use less expensive gases without compromising etching efficiency.

Etching uniformity across the entire substrate surface is among major concerns in microelectronics and diffractive optics because the etch rate can vary in a complicated manner over the surface [52,53,277]. In essence, all recent improvements in plasma etching technology aim to give high etching uniformity and rate; hence the high complexity and cost of the equipment. Our evaluation of off-electrode plasma etching in terms of uniformity showed that both plasma-chemical etching and ion-chemical etching are uniform. To prove this, an experiment was carried out to examine the etching uniformity of a wafer 78 mm in diameter. Etching nonuniformity was obtained from

$$\omega = \frac{\Delta h_m}{h_m} \times 100\%, \qquad (5.64)$$

where h_m is the required height of the microrelief and Δh_m is the deviation of the height from the desired value. Δh_m was measured at test points. Figure 5.31 shows how the test points were arranged on the substrate surface.

Profile measurements taken in different regions of the wafer yielded virtually identical etch depths. The slight nonuniformity observed in the experiment was local and apparently related to the state of the surface itself (its defects, residual contamination, etc.) rather than the plasma. On the whole, the nonuniformity of etching across the wafer was no greater than 1%. The experiment leads to the conclusion that the use of off-electrode plasma is a technique that should prove useful in the fabrication of DOEs and nano- and microelectronic elements.

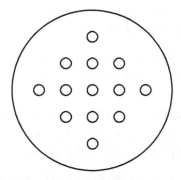

FIGURE 5.31 Arrangement of test points on the substrate surface.

5.8 FABRICATING MICRORELIEFS ON THE SURFACES OF OPTICAL MATERIALS THROUGH PLASMA-CHEMICAL ETCHING IN OFF-ELECTRODE PLASMA

The procedure comprises the following steps:

1. Repeat the sequences of operations described in Sections 2.6, 3.8, and 4.4.
2. Pump the argon out of the working chamber to a pressure of $2 \cdot 10^{-2}$ Torr and fill the chamber with CF_4/O_2 process gas with an oxygen percentage of 0.8%–2% to a pressure of 1 Torr.
3. Repeat step 2.
4. Evacuate the working chamber to a pressure of $2 \cdot 10^{-2}$ Torr, activate the high-voltage gas discharge, and set the discharge current at 80–140 mA (controlled by admitting the process gas through a microleak valve), the electrode voltage at 0.8 kV, and the substrate temperature at 360 K.
5. Etch the substrates for the period required to obtain the desired height of the diffraction microrelief.
6. Deactivate the high-voltage gas discharge and cool the substrates under vacuum conditions for 10 min.
7. Pump CF_4/O_2 out of the vacuum chamber to a pressure of $2 \cdot 10^{-2}$ Torr and fill it with oxygen to a pressure of 1 Torr.
8. Repeat step 7.
9. Evacuate the working chamber to a pressure of $2 \cdot 10^{-2}$ Torr, activate the high-voltage gas discharge, and set the parameters for removing the resist mask: a discharge current of 100 mA (controlled by admitting the process gas through a microleak valve), an accelerating voltage of 2 kV, and a treatment time of 30 s.
10. Repeat step 6.
11. Depressurize the working chamber.
12. Remove the substrates from the holders and cool them in the desiccator for 30 min.
13. Remove the metallized layer by using hydrochloric acid.

14. Wash the substrates in running distilled water.
15. Dry the substrates at 90°C for 30 min.

Steps 7–10 apply if single-layer photoresist masks or double-layer masks (with a metallized sublayer) are used; step 13 applies if a metallized mask is used.

5.9 FABRICATING MICRORELIEFS ON THE SURFACES OF OPTICAL MATERIALS THROUGH ION-CHEMICAL ETCHING IN OFF-ELECTRODE PLASMA

The procedure is mostly similar to that described in Section 5.8, and so we will list only those steps that are relevant to ion-chemical etching:

1. Repeat steps 1–3 described in Section 5.8.
2. Evacuate the working chamber to a pressure of $2 \cdot 10^{-2}$ Torr, activate the high-voltage gas discharge, and set the discharge current at 80–140 mA (controlled by admitting the process gas through a microleak valve), the electrode voltage at 2 kV, and the substrate temperature at 390–440 K.
3. Repeat steps 5–15 described in Section 5.8.

5.10 CHAPTER SUMMARY

The ion-chemical and plasma-chemical techniques for fabricating optical micro-structures 0.5–7.5 μm deep on SiO_2, SiC, and polymer surfaces and diamond-like films would benefit from the use of off-electrode plasma. Techniques involving the use of this plasma are inexpensive, have low power consumption, and produce optical microreliefs with virtually vertical-walled patterns and optically smooth surfaces, as well as completely remove resist masks while keeping the etch nonuniformity to a level as low as 1% across the wafer surface.

The practical recommendations in this chapter can be used for etching materials other than those discussed here with a directed low-temperature plasma flux as well as for describing multicomponent-plasma processes.

6 Generating a Catalytic Mask for the Microrelief of an Optical Element When an Al–Si Structure Is Irradiated by High-Voltage Gas-Discharge Particles

DOEs can be fabricated on silicon substrates through plasma-chemical etching. But this technique produces trenches whose profiles and depths deviate from the design intent [236]. Reference 298 proposes a patented technique for fabricating DOE microreliefs that involves coating a substrate with a catalytic mask. Once the mask is deposited, the structure is heated in a carrier gas by using a large-aperture irradiation flux with a wavelength within the transmission window of the material being treated. In this case, the catalytic mask determines the geometry of microrelief trenches.

This chapter discusses a molten aluminum–silicon system irradiated by a high-voltage gas-discharge ion–electron flux with a particle energy up to 6 keV. This chapter theoretically explores the possibility of atomically sized vacant sites—*vacancies*—existing in the bulk of the liquid phase of aluminum (the mask) when it is exposed to a negatively charged particle flux with an energy up to 6 keV.

This chapter also presents the results of experimental investigations, which agree well with the model of silicon-atom entrainment by a vacancy flux.

We will discuss how catalytic masks are generated for fabricating DOE microreliefs. Such masks are obtained through photolithography techniques by transforming the conventional DOE photomask into a catalytic mask based on molten aluminum. Semiconductor atoms are dispensed into the masking melt by changing the irradiation regimes of the Al–Si structure directly during the process. The subsequent removal of the semiconductor-saturated layer in the catalytic mask produces a diffraction microrelief.

This monograph proposes using the mechanism for dissolving silicon atoms in the aluminum mask to control the parameters of the DOE microrelief being fabricated. This is accomplished during the process by changing the parameters set for irradiating the Al–Si structure with negatively charged particles of high-voltage gas discharge.

6.1 ENTRAINMENT OF SILICON ATOMS BY VACANCIES FORMED IN AN ALUMINUM MELT WHEN ITS SURFACE IS EXPOSED TO HIGH-VOLTAGE GAS-DISCHARGE PARTICLES

In studies relating to mechanisms for dissolving semiconductors in the liquid phase of metals, the problem arises of adjusting the parameters governing the diffusion of semiconductor atoms in the bulk of the melt, such as the concentration of semiconductor atoms in the melt and the process duration. At present, no methods are available for adjusting the melting regimes directly during the process. This problem can be solved with a high-voltage gas discharge serving as an energy source whose distinguishing feature is that it generates fluxes of electrons and negative ions that are almost independent of the gas-discharge device.

A liquid-phase metal can be treated as a highly elongated body that retains the structure of a solid near the melting point [299]. Then, despite its instability, this structure can contain around 1% atomic-size vacancies that have an activation energy on the order of 0.6–0.7 eV [300].

If the average thermal-motion velocity of the liquid-metal atoms is many times lower than the velocity of charged particles in the ion–electron flux [301] and the times taken for a thermal regime to be established in the melt and for irradiation are 0.2 s [87] and 2–20 s, respectively, then a liquid-metal atom can justifiably be considered to be fixed relative to a charged particle, and convection processes are negligible [302]. In this study, both conditions are satisfied, and thus Fick's second equation can be used for calculations [303].

The samples were silicon–aluminum structures consisting of KEF-32 grade silicon and chemically pure aluminum. The structures were irradiated with an ion–electron flux having a particle energy up to 6 keV, a cross-section diameter of 78 mm, and a current up to 200 mA. The process gas was air. The uniformity of the particle-energy distribution over the cross section was better than 98%, and the electron and ion concentrations were $0.3 \cdot 10^{10}$ cm^{-3} [304] and $0.3 \cdot 10^{10}$ cm^{-3}, respectively [305].

During irradiation with an ion–electron flux, two types of particle impinge on the aluminum surface: electrons and negative oxygen ions. The energy that these particles transfer to aluminum atoms is given by the following equations:

For electrons:

$$\Delta T_e = \frac{2 \, e \, U \, m_e}{M}, \tag{6.1}$$

For oxygen ions:

$$\Delta T_i = \frac{4 \, e \, U \, m_i \, M}{(m_i + M)^2}, \tag{6.2}$$

where M is the mass of an aluminum atom; U is the accelerating voltage; m_e is the mass of an electron; and m_i is the mass of an oxygen ion.

Substituting $M = 44.82 \cdot 10^{-27}$ kg and $U = 6$ kV, as well as the electron mass $m_e = 9.1 \cdot 10^{-31}$ kg and the ion mass $m_i = 26.5 \cdot 10^{-27}$ kg into Equations 6.1 and 6.2, we obtain $\Delta T_e = 0.23$ eV and $\Delta T_i = 5.6$ keV, respectively. Because the activation energy of vacancies in liquid metals falls within the range 0.6–0.7 eV [300], which exceeds ΔT_e but is lower than ΔT_i, an ion or several electrons can create in the near-surface layer of molten aluminum a vacancy gradient having a concentration corresponding to the particle concentration in the ion-plasma flux. Taking into account the energies of the electrons and oxygen ions imparted by them to the Al and using References 306 and 307, we can determine the vacancy concentration as $1.5 \cdot 10^{19}$ cm^{-3}. This is higher than the vacancy concentration formed under the action of a purely thermal field ($1.6 \cdot 10^{18}$ cm^{-3}).

When an ion–electron flux irradiates the Al–Si structure, a temperature gradient of 70–35°C forms between them in the 1,053–1,388 K range. The existence of these two gradients leads to the formation of a vacancy flux in the direction of the Si surface. Bearing in mind the vacancy mechanism of diffusion, we can predict that Si atoms will be entrained by the vacancy flux.

If we use only the vacancy-concentration gradient in the Fick's equation when determining the time that Si atoms take to saturate the melt, we observe that in this case the times of saturation for melting in a uniform thermal field (t_1) and under ion–electron irradiation (t_2) are $t_1 = 14 \cdot 10^{-6}$ s and $t_2 = 90$ s. This indicates that in the first case the Si atoms entering the melt reach its surface almost instantaneously. This makes it impossible to control the melting process, which agrees well with known data [232].

In the second case, vacancies entrain the Si atoms, which slows down the diffusion, and the solubility of Si (σ_{Si}) in molten aluminum becomes dependent on the irradiation time (Figure 6.1).

The relationship shows that for irradiation times < 90 s, σ_{Si} is lower than the σ_{Si} derived from the Al–Si state diagram. The maximum difference between these values is observed at $t = 2$ s—that is, when the Si atom front is in the melt, as shown in Figure 6.2a.

From this time onward, it is possible to regulate the quantity of Si in the melt by varying the regime of surface irradiation.

But if entrainment occurs, the solidification of Si in the irradiation zone should follow different laws from that outside that zone. In fact, the angle lap of the melt in Figure 6.2b suggests that the dendritic Si ribbons near the surface in the irradiation zone are predominantly directed almost normally to the Si surface, while at the

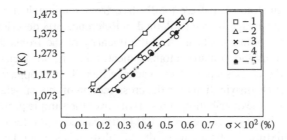

FIGURE 6.1 Variation of Si solubility as a function of melt temperature for various times of ion–electron irradiation: (1) 2 s, (2) 40 s, (3) 60 s, (4) 90 s, and (5) solubility determined from state diagram ($I = 3$–30 mA; $U = 6$ kV).

(a)   (b)

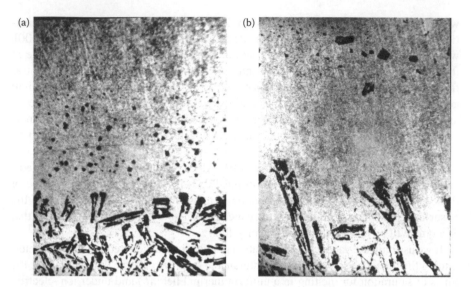

FIGURE 6.2 Distribution of Si atoms in Al after ion–electron irradiation at a current of $I = 2.8$ mA, an accelerating voltage of $U = 3.6$ kV, and an initial Al layer thickness of $h = 3$ mm: (a) beginning of motion of the Si layer dissolved in the melt and distorted by mechanical treatment ($t = 2$ s) and (b) motion time of the layer of Si surface atoms in the middle of the Al layer ($t = 3$ s).

beginning of dissolution (Figure 6.2a), they are distributed randomly. The deviation from the normal is apparently due to the thermal motion that occurs while Al solidifies for a period of 1–2 s.

The Si atoms at the surface are weakly bound to the crystal structure and thus dissolve in liquid metals at a much higher rate. After it enters a vacancy area, a Si atom begins to be reflected from its inner surface, staying in the area for some time, during which the vacancy transfers the atom toward the Si surface. On its path, the Si–vacancy structure encounters both Si atoms and free vacancies. These readily interact to form Si atom complexes and expand the vacancy area—that is, vacancies rapidly stimulate the saturation of the melt with Si atoms. This accounts for the characteristic features of atom entrainment—namely, that it yields an Al melt with maximum Si content and that it can yield melt areas that are totally free from semiconductor atoms (Figure 6.2b). The latter advantage is well explained in Reference 308 describing the mechanism for producing ohmic contacts. And the linearity of the fronts dissolved in the melt of surface atoms and Si matrix atoms indicates that the vacancies are uniformly distributed over the cross section of the vacancy flux. This is apparently due to the uniformity of particle distribution over the cross section of the ion–electron flux.

Figure 6.3 shows the distribution of resistivity over the sample profile. The slightly increasing resistivity in the range $x = 0$–109.15 μm is due to the formation of a layer of solid Si–Al mixture by the atoms of the Si substrate's surface layer. A chemical analysis of the structure showed that this layer could only be removed with a mixture of nitric and hydrofluoric acids, a fact suggesting its semiconducting properties. This was also suggested by a severalfold increase in the amplitude of thermal EMF [301]

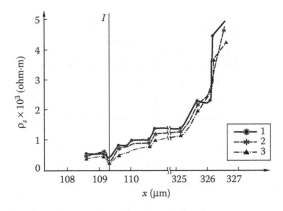

FIGURE 6.3 Distribution of resistivity over the sample profile: (1) 5-element series, (2) 7-element series, and (3) 10-element series. (I) Al–Si interface.

and a change in its sign as compared with the initial aluminum film. The much higher Si content in this area can be explained only by the high concentration of atomic-size vacant sites in which several Si atoms can accumulate and bond to one another.

This process doubtless involves trapping Al atoms, and this leads to the formation of an Al–Si solid solution. At high concentrations of these complexes or even when they merge the vacancy generation ceases, and the layer thickness grows through the diffusion of Si atoms through the liquid Al layer, caused by the temperature gradient in the Al–Si structure and the atoms' thermal motion. Ultimately this may result in the formation of a continuous layer having semiconducting properties. When this layer was removed, we observed a shiny layer of metallic Al ($x = 109.15$–109.3 µm), removable with hot hydrochloric acid. In similar investigations for a Au–Si structure, this layer was gold-colored and could be removed with aqua regia etchant.

Abruptly heating and cooling the structure leads to an exponential increase in the dislocation concentration in the Si crystal from $5 \cdot 10^5$ cm^{-2} at the 327-µm boundary until the dislocations merge at the metal–semiconductor interface ($x = 109.3$ µm) to form dislocation loops and ribbons; the thickness of this layer does not exceed 0.5–0.8 µm. The formation of Cotrell clouds accounts for the low resistivity in the range $x = 109.3$–109.6 µm (Figure 6.3). The dislocations are bulk-type defects as they move in the irradiation zone and can thus transfer Al atoms to appreciable distances ($x = 109.3$–326.6 µm).

To summarize, when an Al–Si structure is bombarded with low-energy particles, an effect involving entrainment of semiconductor atoms by a vacancy flux occurs in the liquid metal, and the parameters of this effect are easily controllable by varying the electron–ion irradiation regimes of the melt surface.

6.2 ANALYTICAL DESCRIPTION OF SILICON DISSOLUTION IN AN ALUMINUM MELT

Reference 91 proposes a physical model for the dissolution of Si atoms in an Al melt when the Al–Si structure is exposed to an ion–electron flux (Figure 6.4).

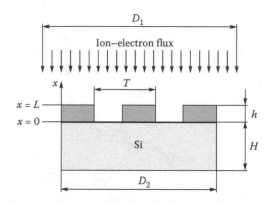

FIGURE 6.4 Irradiation of the Al–Si structure by an ion–electron flux generated by a high-voltage gas-discharge device.

To obtain this structure, the thermal vacuum evaporation method was used to deposit chemically pure aluminum on the surface of a Si substrate with a diameter of $D_2 = 30$ mm and a thickness of $H = 0.3$ mm [309]. Then, through standard photolithographic techniques, a mask was formed on the Si surface, having a periodic lattice with a period of $T = 12$ μm and consisting of aluminum strips with a width of $b = 6$ μm and a thickness of $h = 0.2$ μm. For metallographic examinations, structures with a 4-μm-thick aluminum layer were used as well.

Note that the following discussion holds only for $h \ll b$, where b is the mask-groove width. Satisfaction of this inequality as well as the action of surface tension at the molten aluminum–silicon interface prevent the molten aluminum spreading to other areas of the semiconductor during irradiation.

Section 6.1 explained how a vacancy flux (j_v) directed toward a semiconductor surface is formed.

It is known [302] that in a molten state almost all substances in equilibrium systems and homogeneous systems have approximately the same diffusivity, $D \approx 5 \cdot 10^{-4}$ cm²/s. In nonequilibrium environments, its value depends on the concentration and activation energy of vacancies. For determining the spread rate of vacancies (V), the diffusivity can be assumed to be constant and equal to $D_v \approx 5 \cdot 10^{-4}$ cm²/s. Because $V \cdot \lambda_v \approx D_v$ (λ_v is the mean free path of vacancies in the melt), with a mask thickness of $h = 4$ μm or 0.2 μm, vacancies will reach the surface of the Al–Si interface almost instantaneously. The variation of the vacancy flux in the bulk of molten aluminum can be expressed as the Fick's equation [232]

$$j_v = D_v \left(\frac{\partial C_v}{\partial x} \right), \tag{6.3}$$

where D_v is the diffusivity of vacancies and C_v is the concentration of vacancies.

Silicon dissolves in aluminum under the vacancy mechanism [299], whereby those vacancies that have reached the Si surface stimulate the formation of a counterflux of Si atoms (j) that is directed from the Si surface to the Al surface.

In moving toward the Al surface, the Si atoms will interact with nonuniformly distributed vacancies. In this case, the Si diffusivity will show a functional dependence on the vacancy concentration. The diffusivity can be written as

$$D(C_v) = \left(\frac{C_v}{C_0} \right) D_v, \tag{6.4}$$

where $D(C_v)$ is the Si diffusivity in molten aluminum and C_0 is the maximum concentration of Si atoms in the aluminum melt.

Given this condition, the counterflux of Si atoms moving toward the Al surface can be described by the expression

$$j = D(C_v) \left(\frac{\partial C}{\partial x} \right), \tag{6.5}$$

where C is the concentration of Si atoms in the melt.

By substituting Equation 6.4 into Equation 6.5 and then Equations 6.3 and 6.5 into the continuity equation

$$\frac{\partial C}{\partial t} = -\text{div}(\vec{j}), \tag{6.6}$$

we obtain a system of equations that describes silicon dissolution in aluminum in the presence of a vacancy flux,

$$\frac{\partial C_v}{\partial t} = D_v \frac{\partial^2 C_v}{\partial x^2};$$

$$\frac{\partial C}{\partial t} = \frac{D_v}{C_0} \frac{\partial}{\partial x} \left(C_v \frac{\partial C}{\partial x} \right). \tag{6.7}$$

This system of equations takes into account the experimentally ascertained absence of a reverse process (dissolution of Al atoms in Si)—that is, the presence of the Al–Si interface during irradiation. At the same time, experiments have shown that the Al–Si interface remains at the same level ($x = 0$; see Figure 6.4) until aluminum reaches a molten state.

From this time and until the irradiation is complete, silicon dissolves layer by layer in molten aluminum. As a result, Al layers saturated with semiconductor atoms replace Si layers and thus cause the level of $x = 0$ to shift by a value of h_l. Experiments have proved that the boundary $x = L$ also shifts by a value of h_l toward the semiconductor surface. Indeed, as Figure 6.5a–c shows, the height of the masking material before irradiation was equal to the thickness of the deposited aluminum layer and after irradiation it dropped to zero.

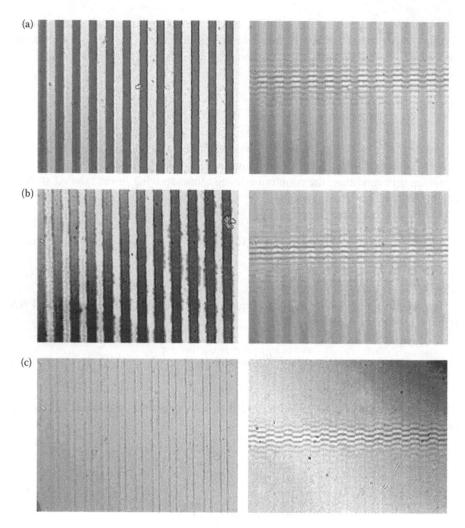

FIGURE 6.5   Al–Si structures and related interferograms: (a) before irradiation, (b) after irradiation, and (c) substrate surface after removal of the catalytic mask.

But when the semiconductor-saturated layer of the catalytic mask was removed, the substrate's surface was found to have trenches with a height of h_l, equal to the thickness of the deposited aluminum layer (see Figure 6.5a and c). Thus, the levels $x = 0$ and $x = L$ doubtlessly shift toward the semiconductor, but we can safely assume that the value by which they shift remains unchanged and equal to h_l throughout the irradiation.

Silicon dissolution in molten aluminum under the vacancy mechanism assumes the presence of constant sources [309]: those of vacancies at the melt surface ($x = L$) and of semiconductor atoms at the boundary $x = 0$. Either a thermal field or, as noted earlier, an ion–electron flux can serve as a source of vacancies. In either case, the formation of vacancies at the melt surface free from foreign semiconductor atoms precedes the dissolution. But creating a gradient of vacancies at the boundary $x = 0$

is impossible because the number of Si surface atoms is many times greater than the number of vacancies that have reached the semiconductor surface. This results in the Si surface atoms absorbing the vacancies. Si atoms saturate the melt in the direction from the semiconductor surface to the melt surface. Because the concentration of Si atoms in the melt depends on its solubility limit [310], creating a gradient of Si atoms at the boundary $x = L$ is not possible either. Given these considerations, the terminal and initial conditions for the system of equations (6.7) are as follows:

$$C_v\big|_{x=L} = C_{v0}; \quad \frac{\partial C_v}{\partial x}\bigg|_{x=0} = 0; \quad C_v\big|_{t=0} = \begin{cases} C_{v0} & \text{for } x = L \\ 0 & \text{for } 0 \le x < L \end{cases}.$$

$$C\big|_{x=0} = C_0; \quad \frac{\partial C}{\partial x}\bigg|_{x=L} = 0; \quad C\big|_{t=0} = \begin{cases} C_0 & \text{for } x = 0 \\ 0 & \text{for } 0 < x \le L \end{cases}. \tag{6.8}$$

The first and the fourth terminal conditions (from left to right, line by line) in Equation 6.8 indicate that constant sources of vacancies at the melt surface and of semiconductor atoms at the boundary $x = 0$ are present. The second and the fifth conditions of the second order show that the concentrations of vacancies at $x = 0$ and of semiconductor atoms at $x = L$ remain unchanged—that is—that the gradients of both the vacancies and the semiconductor atoms here equal zero. The third and the sixth initial conditions highlight that at $t = 0$ the bulk of the melt contains neither foreign semiconductor atoms nor vacancies.

6.2.1 CONSERVATIVE DIFFERENCE SCHEME FOR DIFFUSION EQUATIONS

Let us write the nonlinear homogeneous equation for diffusion as [311]

$$\frac{\partial y}{\partial t} = \frac{\partial}{\partial x}\left[K(x)\frac{\partial y}{\partial x}\right], \tag{6.9}$$

where $K(x)$ is diffusivity.

Then, let us introduce the uniform grid

$$\omega_{\tau h} = \{t^i = i\tau, x_n = nh, i = \overline{0, N_1}, n = \overline{2, N_2}\},$$

on which we can write the difference equations

$$\frac{1}{\tau}(y_n^{i+1} - y_n^i) = \frac{1}{h^2}\left[x_{n+\frac{1}{2}}(y_{n+1}^{i+1} - y_n^{i+1}) - x_{n-\frac{1}{2}}(y_n^{i+1} - y_{n-1}^{i+1})\right], \tag{6.10}$$

where

$$x_{n+\frac{1}{2}} = \frac{1}{2}[K(x_n) + K(x_{n+1})];$$

$$x_{n-\frac{1}{2}} = \frac{1}{2}[K(x_{n-1}) + K(x_n)].$$

This scheme is conservative and therefore absolutely stable [312]. It has the first order of approximation in time and second order in space [312].

Similarly, we can construct a difference scheme for the system of equations (6.7):

$$\frac{1}{\tau}(C_{vn}^{i+1} - C_{vn}^{i}) = \frac{D_v}{h^2}[C_{vn+1}^{i+1} - 2C_{vn}^{i+1} + C_{vn-1}^{i+1}]; \qquad (6.11)$$

$$\frac{1}{\tau}(C_n^{i+1} - C_n^{i}) = \frac{1}{2}\frac{D_v}{C_0 h^2}((C_{vn+1}^{i+1} + C_{vn}^{i+1})(C_{n+1}^{i+1} - C_n^{i+1})$$
$$- (C_{vn}^{i+1} + C_{vn-1}^{i+1})(C_n^{i+1} - C_{n-1}^{i+1})).$$

For this system of equations, the following difference approximations obtained from Equation 6.8 for terminal and initial conditions hold:

$$C_{v2}^{i} = C_{v1}^{i}, \quad i = \overline{0, N_1};$$

$$C_{vN_2}^{i} = C_{v0}, \quad i = \overline{0, N_1};$$

$$C_{N_2-1}^{i} = C_{N_2}^{i}, \quad i = \overline{0, N_1};$$

$$C_1^{i} = C_0, \quad i = \overline{0, N_1};$$

$$C_{vn}^{0} = \begin{cases} C_{v0}, & n = N_2, \\ 0, & 1 \leq n < N_2; \end{cases} \quad C_n^{0} = \begin{cases} C, & n = 1, \\ 0, & 1 < n \leq N_2 \end{cases}. \qquad (6.12)$$

6.2.2 Difference Solution to the Mixed Problem

Difference equations (6.11) were solved with the sweep method [313]. The results are presented in Figure 6.6a–c [309,314,315].

Similar results are obtainable if we assume that there is a certain vacancy gradient (A) in the melt and that its distribution over the melt's cross section is described by the function $C_v = Ax + B$. This holds true for low values of the gradient (A) when the function *erf* describing the concentration profiles [232] is approximated well by a straight line at short segments of the curve $C = f(x, t)$. In this study, this condition is satisfied because creating significant temperature gradients in molten aluminum and, therefore, vacancy gradients in melts within a length of 3–4 μm is almost impossible: the difference of the vacancy concentration at the Al surface $(x = L)$ and

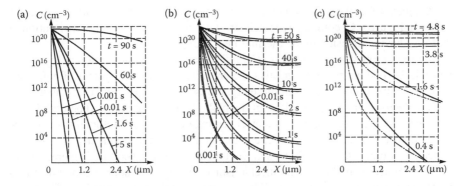

FIGURE 6.6 Distribution of Si atom concentration over the cross section of the aluminum melt: (a) $C_{V0} = 10^{18}$ cm^{-3} ($A = 0$), (b) $C_{V0} = 3 \cdot 10^{18}$ cm^{-3} ($A = 2 \cdot 10^{18}$), and (c) $C_{V0} = 5 \cdot 10^{18}$ cm^{-3} ($A = 4 \cdot 10^{18}$). For all cases, $D_v = 5 \cdot 10^{-4}$ cm^2/s. Continuous lines describe the solution to the system of equations (6.7); dot-and-dash lines, the solution to Equation 6.14; and X stands for the experimental values of Si atom concentration measured with a MAR-2 microanalyzer.

that at the Si surface ($x = 0$; see Figure 6.4) is less than one order of magnitude. Thus, the diffusivity of semiconductor atoms in the melt is given by

$$D(C_v) = \frac{Ax + B}{C_0} \cdot 5 \cdot 10^{-4}. \tag{6.13}$$

Substituting this expression into the second equation in Equation 6.7, we obtain an expression that describes the change in the concentration of semiconductor atoms in the melt:

$$\frac{\partial C}{\partial t} = \left[A \frac{\partial C}{\partial x} + (Ax + B) \frac{\partial^2 C}{\partial x^2} \right] \cdot \frac{5 \cdot 10^{-4}}{C_0}. \tag{6.14}$$

Figure 6.6a–c presents a numerical solution to Equation 6.14 for initial and terminal conditions (6.8); it is shown as curves characterizing the distribution of Si atom concentration over the cross section of the aluminum melt.

6.2.3 ANALYSIS OF NUMERICAL RESULTS

In the absence of the vacancy flux observed if $C_{V0} = C'_{V0} = 10^{18}$ cm^{-3}, where C'_{V0} is the vacancy concentration at the melt surface under the action of the thermal field, semiconductor atoms have an equal opportunity to move in any direction in the melt—that is, this process is similar to diffusion in the thermal field. In this case, the distribution of Si atom concentration over the cross section of the melt is described by the Fick's equation

$$\frac{\partial C}{\partial t} = D \frac{\partial^2 C}{\partial x^2}. \tag{6.15}$$

This means that the diffusion in the melt is determined only by dC/dx and that the curves in Figure 6.6a completely match similar curves describing the distribution of Si atoms during diffusion in the thermal field [303].

The formation of a vacancy flux ($C_{v0} > C'_{v0}$) causes a vacancy gradient to form in the melt. Diffusion processes in this case are described by the system of equations (6.7) and expression (6.14). From these equations, we determined the distribution of semiconductor-atom concentrations in the melt (Figure 6.6b and c). The high initial steepness of the curves indicates a slower removal of semiconductor atoms from the semiconductor surface. This is because the semiconductor atoms are entrained by a vacancy counterflux [91] when each Si atom is under the action of several vacancies. As Figure 6.6b shows, this mechanism is strongest at short irradiation times ($t < 0.1$ s): the melt is not in an equilibrium state and there is a gradient of vacancy concentration across its bulk. With an increase in the irradiation time ($t > 0.1$ s), the melt begins to attain an equilibrium state, and the vacancy effect on the stimulation of diffusion intensifies.

As Figure 6.6b and c shows, the diffusion is notably accelerated already at $t > 1$ s. Thus, in addition to the entrainment, another process is observed—that of stimulating the diffusion of foreign atoms because of an increased vacancy concentration. It is this stimulation that accounts for the reduction in the time that Si atoms take to saturate the melt: the time decreases from $t = 90$ s (Figure 6.6a) to $t = 4.8$ s (Figure 6.6c). A feature of the saturation curve is a decrease in C in the area of the semiconductor surface (see Figure 6.6b and c). This could be because the solubility limit of the semiconductor in a melt is invariably lower than the surface concentration of the semiconductor. Otherwise the melt would cease to exist.

6.3   ANALYSIS OF EXPERIMENTAL DATA

To prove the stimulation of Si atom diffusion in molten aluminum, we carried out an experiment to compare the depths up to which aluminum is doped with silicon when the Al–Si structure is exposed to a purely thermal field and ion–electron irradiation. For this purpose, an aluminum layer ($h = 4$ μm) was formed on both surfaces of a silicon wafer. The upper surface of the Al–Si–Al structure was irradiated with an ion–electron flux at a current of $I = 6.9$ mA and an accelerating voltage of $U = 3.8$ kV for a period of $t = 2$ s, while the lower surface was exposed only to the action of the thermal field. Doping depths were metallographically measured with the angle-lap technique ($\alpha = 7°$) [232]. Doped layers were dyed in a mixture of 0.05%–0.1% of 70% nitric acid and 48% hydrofluoric acid. An angle-lap image was obtained with an MII-4 optical microscope. The angle lap is shown in Figure 6.7.

Although both surfaces were at almost the same temperature, 1,123 K, the doping area on the exposed side (the upper part with a height of h_d in Figure 6.7) is greater than that on the unexposed side almost by a factor of three. This proves that during ion–electron irradiation, diffusion is stimulated. In this case, as the curves in Figure 6.6b and c show, varying the irradiation parameters enables controlled

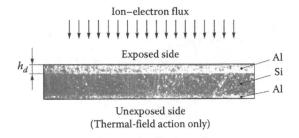

FIGURE 6.7   Angle lap of the Al–Si–Al structure.

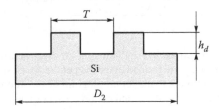

FIGURE 6.8 Parameters of the Si wafer microrelief obtained by irradiating the Al–Si structure with high-voltage gas-discharge particles. The trenches are formed in the area of the Al–Si structure.

change of Si atom concentration in the melt and, therefore, of the depth (h_d) up to which the near-surface area of the semiconductor is doped with the masking material. This means that in the case of a DOE microrelief, the microrelief's depth depends on h_d (see Figure 6.8).

With these results in mind, we set ourselves a task to produce on a silicon substrate a diffraction microrelief with a height of $h_m = 0.2 \, \mu m$ and a period of $T = 12 \, \mu m$ by using a catalytic mask. For this purpose, we irradiated the structure shown in Figure 6.4 with an ion–electron flux generated by a high-voltage gas discharge with a discharge current of $I = 80 \, mA$ and an accelerating voltage of $U = 1 \, kV$. The irradiation time t was 1 min and the Al mask thickness h, 0.2 μm. The discharge current and accelerating voltage were selected such that the Al–Si structure could be heated to a temperature sufficient for the aluminum to melt. Another factor taken into account was that the accelerating voltage must not exceed the critical value at which the energy of negative oxygen ions would have been sufficient for sputtering the silicon [209]. After the semiconductor-saturated Al layer was irradiated and then removed with aqua regia etchant [91], the substrate surface had a diffraction microrelief on it (see Figure 6.9).

As Figure 6.9 shows, the microrelief parameters are consistent with those desired. The cause of the deviation from the normal is that when vacancies impinge on the profile surface, the surface absorbs the vacancies, thereby causing their concentration to diminish, and this, in turn, reduces the bottom width of the microrelief profile.

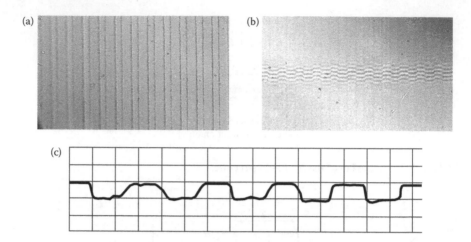

FIGURE 6.9 Microrelief obtained by irradiating the Al–Si structure with high-voltage gas-discharge particles (the trenches are formed in the area of the Al–Si structure): (a) microrelief, (b) interferogram of the microrelief, and (c) profilogram of the microrelief. One division on the horizontal scale is equivalent to 4 µm and on the vertical scale to 0.2 µm.

6.4   FABRICATING A MICRORELIEF BASED ON A CATALYTIC MASK FORMED IN OFF-ELECTRODE PLASMA

The procedure consists of the following steps:

1. Repeat the sequence of steps described in Sections 2.6 and 3.8.
2. By using the techniques of vacuum or chemical deposition, coat each substrate with an aluminum layer with a thickness that satisfies the condition $h \ll b$ and that is equal to the desired height of the diffraction microrelief.
3. Repeat steps 3–8 described in Section 4.4.
4. Etch the aluminum in a selective etchant until the aluminum is completely removed.
5. Remove the photoresist layer with chemically pure acetone.
6. Repeat steps 10–12 described in Section 4.4.
7. Evacuate the working chamber to a pressure of $2 \cdot 10^{-1}$ Torr, activate the high-voltage gas discharge, and set the discharge current at 80–140 mA (controlled by admitting the process gas through a microleak valve) and the electrode voltage at 1–1.2 kV.
8. Irradiate the substrates for the period required to obtain the desired height of the diffraction microrelief.
9. Deactivate the high-voltage gas discharge and cool the substrates under vacuum conditions for 10 min.
10. Depressurize the working chamber.
11. Remove the substrates from the holder and cool them in the desiccator for 30 min.
12. Remove the semiconductor-saturated Al layer with aqua regia etchant.

13. Wash the substrates in running distilled water.
14. Dry the substrates at 90°C for 30 min.

6.5   CHAPTER SUMMARY

Standard plasma-chemical etching techniques may cause the profile and depth of a microrelief trench to deform. To resolve this problem, this chapter proposes and experimentally investigates an off-electrode plasma technique for fabricating microreliefs that involves depositing a molten-aluminum catalytic mask on the surface of a silicon substrate and irradiating the surface with high-voltage gas-discharge particles. This technique makes it possible to control microrelief parameters by changing the mask layout and the irradiation regimes.

The investigations described in this chapter present the conclusion that adopting the proposed approach would also benefit the fabrication of microreliefs on materials other than those discussed here.

Conclusion

The study cycle described in this monograph attests to the promising prospects for the line of scientific work that relates to using new off-electrode plasma techniques and equipment for fabricating optical microreliefs and micro- and nanostructures. This new equipment differs from conventional equipment in that it is both simple and versatile. And the new techniques are inexpensive and energy-efficient and free from the drawbacks inherent in equivalent techniques used in and outside Russia.

These new techniques and equipment help solve a major problem in diffractive optics: how to implement a series of off-electrode plasma processes for fabricating diffraction microstructures on large-aperture wafers.

This monograph experimentally proves how efficient off-electrode plasma is for cleaning substrate surfaces, for enhancing the adhesion of thin metal films, and for generating catalytic masks to fabricate microreliefs, as well as for plasma-chemical and ion-chemical etching.

We have obtained scientific data on newly discovered physical phenomena and effects observed at the surface and in the bulk of a solid when it is exposed to off-electrode plasma. These data help expand the existing theoretical knowledge about processes involved in interaction between low-temperature plasma and the surface.

The results presented here provide a basis for developing a comprehensive solution to the problem of fabricating diffraction microstructures on large-format wafers and for broadening the DOE range with the prospects of diffractive optics and computer diffractive optics in mind.

The results also stimulate new developments in gas discharge physics, plasma physics, solid-state physics, and diffractive and computer diffractive optics. This encourages us to continue our work by carrying out fundamental research into the following problems:

- Diagnosing off-electrode plasma in detail—for example, through modern techniques using the Langmuir probe—and interpreting data obtained
- Elaborating physical and mathematical models illustrating interaction between off-electrode plasma particles and the surfaces of solids, hetero-structures, and polymers, by interpreting the results obtained and physical phenomena discovered
- Retrofitting existing reactors and devices
- Creating new devices and studying their electrical and physical properties and the off-electrode plasma generated with these devices
- Changing over to using off-electrode plasma for fabricating nanostructures and to using technology based on the proposed off-electrode plasma techniques for fabricating microreliefs on the surfaces of optical materials

Appendix A: Statistical Analysis of Experimental Results

TABLE A.1

Average Voltages Supplied to the Electrodes of the Gas-Discharge Device for Different Discharge Currents at a Working-Chamber Pressure of $1.5 \cdot 10^{-1}$ Torr

No.	Discharge Current (mA)	Average Voltage (V)	Error (%)
1	0	298.0 ± 26.48	9.0
2	8	393.5 ± 21.58	5.0
3	10	452.0 ± 30.84	7.0
4	20	594.0 ± 30.73	5.0
5	50	696.8 ± 25.63	4.0
6	100	$1{,}000.5 \pm 22.87$	2.0

TABLE A.2

Average Voltages Supplied to the Electrodes of the Gas-Discharge Device for Different Discharge Currents at a Working-Chamber Pressure of $1.2 \cdot 10^{-1}$ Torr

No.	Discharge Current (mA)	Average Voltage (V)	Error (%)
1	0	298.0 ± 29.52	10.0
2	10	493.0 ± 28.40	6.0
3	15	602.0 ± 20.44	3.0
4	20	647.0 ± 29.95	5.0
5	40	798.0 ± 20.44	3.0
6	60	953.0 ± 28.60	3.0
7	80	$1{,}106.0 \pm 27.62$	2.0
8	110	$1{,}452.0 \pm 31.21$	2.0

TABLE A.3

Average Voltages Supplied to the Electrodes of the Gas-Discharge Device for Different Discharge Currents at a Working-Chamber Pressure of $9 \cdot 10^{-2}$ Torr

No.	Discharge Current (mA)	Average Voltage (V)	Error (%)
1	0	303.5 ± 18.46	6.0
2	8	503.0 ± 19.66	4.0
3	12	663.0 ± 26.32	4.0
4	20	793.0 ± 24.77	3.0
5	55	$1,055.0 \pm 28.44$	3.0
6	72	$1,153.0 \pm 34.37$	3.0
7	100	$1,501.0 \pm 30.41$	2.0
8	110	$1,649.0 \pm 31.50$	2.0

TABLE A.4

Average Instrument Readings for Different Exposure Times of Substrates in the Contaminator Placed into the Vacuum Chamber, with the Indenter Inclined at an Angle of $\beta = 5.5$ and the Substrate Holder at an Angle of $\alpha = 60°$

No.	Exposure Time (min)	Average Pulse Time (ms)	Error (%)
1	0	9.99 ± 0.198	2.0
2	60	0.499 ± 0.0198	4.0

TABLE A.5

Average Pulse Times τ for Different Indenter Angles β at $\alpha = 60°$ for a Substrate Series with the Initial Contamination Level $\tau = 0.7 \cdot 10^{-2}$ s

No.	β (°)	Average Pulse Time (ms)	Error (%)
1	4	7.065 ± 0.1092	2.0
2	5.5	6.74 ± 0.057	1.0
3	6	7.147 ± 0.1074	2.0
4	6.5	9.471 ± 0.2550	3.0

TABLE A.6
Average Pulse Times τ for Different Indenter Angles β at $\alpha = 60°$ for a Substrate Series with the Initial Contamination Level $\tau = 0.4 \cdot 10^{-2}$ s

No.	β (°)	Average Pulse Time (ms)	Error (%)
1	4	4.091 ± 0.1017	2.0
2	5.5	2.986 ± 0.0302	1.0
3	6	3.399 ± 0.0647	2.0
4	6.5	7.13 ± 0.212	3.0

TABLE A.7
Average Pulse Times τ for Different Angles of the Test-Substrate Holder at $\beta = 5.5$ for a Substrate Series with the Initial Contamination Level $\tau = 0.9 \cdot 10^{-2}$ s

No.	α (°)	Average Pulse Time (ms)	Error (%)
1	40	9.029 ± 0.1215	1.0
2	55	8.035 ± 0.1198	1.0
3	60	7.058 ± 0.1778	3.0

TABLE A.8
Average Pulse Times τ for Different Angles of the Test-Substrate Holder at $\beta = 5.5°$ for a Substrate Series with the Initial Contamination Level $\tau = 0.45 \cdot 10^{-2}$ s

No.	α (°)	Average Pulse Time (ms)	Error (%)
1	30	4.532 ± 0.0650	1.0
2	55	3.054 ± 0.0407	1.0
3	60	2.394 ± 0.0718	3.0

TABLE A.9
Average Pulse Times τ Corresponding to Different Numbers of
Passes along a Single Path for Three Batches of Differently
Contaminated Substrates from a Substrate Series with the Initial
Contamination Level τ = 0.5 · 10^{-2} s

No.	n	Average Pulse Time (ms)	Error (%)
1	1	5.011 ± 0.1192	2.0
2	4	7.339 ± 0.1359	2.0
3	5	6.893 ± 0.2319	3.0
4	7	8.391 ± 0.2525	3.0
5	8	9.036 ± 0.3388	4.0
6	9	9.906 ± 0.3725	4.0
7	10	11.864 ± 0.5505	5.0

TABLE A.10
Average Pulse Times τ Corresponding to Different Numbers of
Passes along a Single Path for Three Batches of Differently
Contaminated Substrates from a Substrate Series with the Initial
Contamination Level τ = 0.3 · 10^{-2} s

No.	n	Average Pulse Time (ms)	Error (%)
1	1	3.006 ± 0.0981	3.0
2	3	3.733 ± 0.1301	3.0
3	5	4.221 ± 0.1756	4.0
4	7	6.481 ± 0.2411	4.0
5	8	6.361 ± 0.2854	4.0
6	9	6.605 ± 0.2979	5.0
7	11	8.829 ± 0.4963	6.0

TABLE A.11
Average Pulse Times τ Corresponding to Different Numbers of Passes along a Single Path for Three Batches of Differently Contaminated Substrates from a Substrate Series with the Initial Contamination Level $\tau = 0.1 \cdot 10^{-2}$ s

No.	n	Average Pulse Times (ms)	Error (%)
1	1	1.046 ± 0.0811	8.0
2	3	1.117 ± 0.0508	5.0
3	5	2.394 ± 0.1203	5.0
4	7	3.566 ± 0.2092	6.0
5	9	6.177 ± 0.3705	6.0
6	11	8.008 ± 0.5527	7.0
7	13	9.223 ± 0.3732	4.0
8	14	9.577 ± 0.2818	3.0

TABLE A.12
Average Pulse Times τ for Different Numbers of Measurements Taken to Measure the Surface Cleanliness of Test Substrates with One Indenter for a Series of Substrates with the Initial Contamination Level $\tau = 0.4 \cdot 10^{-2}$ s

No.	n	Average Pulse Time (ms)	Error (%)
1	1	4.053 ± 0.0952	2.0
2	3	4.086 ± 0.0100	2.0
3	5	4.391 ± 0.1455	3.0
4	6	4.116 ± 0.0909	2.0
5	7	4.507 ± 0.1332	3.0
6	9	4.548 ± 0.1884	4.0
7	10	4.588 ± 0.2430	5.0

TABLE A.13
Average Pulse Times τ for Different Numbers of Measurements Taken to Measure the Surface Cleanliness of Test Substrates with One Indenter for a Series of Substrates with the Initial Contamination Level $\tau = 0.6 \cdot 10^{-2}$ s

No.	n	Average Pulse Time (ms)	Error (%)
1	1	6.040 ± 0.1032	2.0
2	4	7.117 ± 0.3140	4.0
3	5	7.659 ± 0.4519	6.0
4	6	8.174 ± 0.4726	6.0

TABLE A.14
Average Pulse Times τ for Different Numbers of Measurements Taken to Measure the Surface Cleanliness of Test Substrates with One Indenter for a Series of Substrates with the Initial Contamination Level $\tau = 0.8 \cdot 10^{-2}$ s

No.	n	Average Pulse Time (ms)	Error (%)
1	1	8.080 ± 0.1918	2.0
2	2	8.757 ± 0.2343	3.0
3	3	9.892 ± 0.4743	5.0

TABLE A.15
Average Pulse Times τ for Different Air-Exposure Times for a Series of Substrates with the Initial Contamination Level $\tau = 0.2 \cdot 10^{-2}$ s

No.	Air-Exposure Time (min)	Average Pulse Time (ms)	Error (%)
1	1	2.053 ± 0.0734	4.0
2	2	1.043 ± 0.1184	11.0

TABLE A.16

Average Pulse Times τ for Different Air-Exposure Times for a Series of Substrates with the Initial Contamination level $\tau = 0.5 \cdot 10^{-2}$ s

No.	Air-Exposure Time (min)	Average Pulse Time (ms)	Error (%)
1	1	5.002 ± 0.1026	2.0
2	4	3.150 ± 0.1499	5.0

TABLE A.17

Average Pulse Times τ for Different Air-Exposure Times for a Series of Substrates with the Initial Contamination level $\tau = 0.9 \cdot 10^{-2}$ s

No.	Air-Exposure Time (min)	Average Pulse Time (ms)	Error (%)
1	1	9.022 ± 0.1153	1.0
2	4	6.654 ± 0.1866	3.0
3	5	6.426 ± 0.2837	4.0

TABLE A.18

Average Copper Film–Polycrystalline Substrate Adhesion for Different Contamination Times for the Untreated Structure

No.	Contamination Time (min)	Average Adhesion Strength (N/mm^2)	Error (%)
1	0	8.5 ± 1.00	11.8
2	1	10.0 ± 1.0	10.0
3	2	14.0 ± 0.52	3.7
4	3	16.0 ± 0.54	3.4
5	5	11.5 ± 0.51	4.4
6	7	8.0 ± 0.253	3.2
7	10	4.25 ± 0.251	5.9

TABLE A.19

Average Copper Film–Polycrystalline Substrate Adhesion for Different Contamination Times for the Structure Treated at $I = 100$ mA, $U = 2$ kV, and $t_{ir} = 5$ min

No.	Contamination Time (min)	Average Adhesion Strength (N/mm²)	Error (%)
1	0	15.15 ± 0.516	3.0
2	1	15.95 ± 0.430	3.0
3	2	16.88 ± 0.646	4.0
4	3	17.00 ± 0.655	4.0
5	7	11.91 ± 0.401	3.0
6	10	11.27 ± 0.269	2.0

TABLE A.20

Average Copper Film–Polycrystalline Substrate Adhesion for Different Bombardment Times of the Me–C_xH_y–Sub Structure at $I = 100$ mA, $U = 2$ kV, and $t_{cont} = 10$ min

No.	Exposure Time (min)	Average Adhesion Strength (N/mm²)	Error (%)
1	0	4.5 ± 0.51	11.3
2	0.5	5.1 ± 0.65	12.7
3	1	6.3 ± 0.80	12.6
4	1.5	8.2 ± 0.55	6.7
5	2	11.5 ± 0.75	6.5
6	2.5	15.5 ± 0.50	3.2
7	3	18.25 ± 1.000	5.5
8	3.5	18.75 ± 0.750	4.0
9	4	19.5 ± 1.00	5.1
10	4.5	19.75 ± 1.250	6.3
11	5	20.0 ± 1.5	7.5

TABLE A.21
Average Copper Film–Polycrystalline Substrate Adhesion for Different Discharge Currents at $U = 2$ kV, $t_{ir} = 3$ min, and $t_{cont} = 10$ min

No.	Discharge Current (mA)	Average Adhesion Strength (N/mm^2)	Error (%)
1	0	4.5 ± 0.50	11.1
2	10	5.25 ± 0.51	9.7
3	20	6 ± 0.75	12.5
4	30	7 ± 0.5	7.1
5	40	9 ± 0.75	8.3
6	50	12 ± 1.0	8.3
7	60	15.5 ± 1.00	6.5
8	70	16.75 ± 2.000	11.9
9	80	17.5 ± 2.00	11.4
10	90	18 ± 1.0	5.6
11	100	18 ± 0.75	4.2

TABLE A.22
Average Copper Film–Polycrystalline Substrate Adhesion for Different Accelerating Voltages at $I = 20$ mA, $t_{ir} = 3$ min, and $t_{cont} = 10$ min

No.	Voltage (kV)	Average Adhesion Strength (N/mm^2)	Error (%)
1	0	4.5 ± 0.41	9.1
2	1	4.75 ± 0.250	5.3
3	1.5	4.85 ± 0.250	5.2
4	2	6.4 ± 0.62	9.7
5	2.5	10.0 ± 1.1	11
6	3	14.5 ± 1.50	10.3
7	3.5	16.0 ± 1.0	6.25
8	4	17.0 ± 1.1	6.5
9	4.5	17.5 ± 1.250	7.1
10	5	17.6 ± 0.90	5.1

TABLE A.23

Average Etch Rates in CF₄ Plasma for Different Electrode Voltages of the Gas-Discharge Device for a Discharge Current of 140 mA

No.	Voltage (kV)	Average Etch Rate (Å/min)	Error (%)
1	0.8	147.7 ± 15.25	10.0
2	1.5	196.0 ± 15.1	8.0
3	1.7	404.5 ± 23.00	6.0
4	1.8	899.5 ± 24.44	3.0
5	2	1,189.0 ± 74.5	6.5
6	2.3	771.0 ± 55.8	7.0
7	2.5	508.0 ± 51.6	10.0
8	2.7	303.5 ± 28.98	10.0

TABLE A.24

Average Etch Rates in CF₄ Plasma for Different Electrode Voltages of the Gas-Discharge Device for a Discharge Current of 120 mA

No.	Voltage (kV)	Average Etch Rate (Å/min)	Error (%)
1	0.8	50.0 ± 4.1	8.0
2	1.5	146.5 ± 15.23	10.0
3	1.7	304.5 ± 12.43	4.0
4	2	625.5 ± 21.27	3.0
5	2.3	404.5 ± 21.27	5.0
6	2.5	246.0 ± 23.6	10.0
7	2.7	204.0 ± 10.1	5.0

TABLE A.25

Average Etch Rates in CF₄ Plasma for Different Electrode Voltages of the Gas-Discharge Device for a Discharge Current of 80 mA

No.	Voltage (kV)	Average Etch Rate (Å/min)	Error (%)
1	1.5	145.5 ± 10.58	7.0
2	1.7	251.5 ± 11.68	5.0
3	2	406.0 ± 15.01	4.0
4	2.3	300.0 ± 22.8	8.0
5	2.5	203.5 ± 14.40	7.0
6	2.7	138.0 ± 13.1	9.0
7	3	108.5 ± 8.76	8.0
8	3.5	47.0 ± 4.8	10.0

TABLE A.26

Average Etch Rates in CF$_4$ Plasma for Different Electrode Voltages of the Gas-Discharge Device for a Discharge Current of 50 mA

No.	Voltage (kV)	Average Etch Rate (Å/min)	Error (%)
1	1.5	82.0 ± 7.6	9.0
2	2	272.0 ± 13.1	5.0
3	2.3	206.5 ± 12.27	6.0
4	2.7	100.5 ± 10.31	10.0
5	3	75.5 ± 7.24	10.0
6	3.5	30.3 ± 3.18	10.0

TABLE A.27

Average Plasma-Chemical Etch Rates for Different Oxygen Percentages in CF$_4$/O$_2$ Plasma Generated by a High-Voltage Gas Discharge at a Discharge Current of 50 mA (Electrode Voltage $U = 0.8$ kV)

No.	Oxygen Percentage (%)	Average Etch Rate (nm/min)	Error (%)
1	0.2	5.1 ± 0.53	10.0
2	0.5	9.9 + 0.86	9.0
3	3.0	7.1 ± 0.71	10.0

TABLE A.28

Average Plasma-Chemical Etch Rates for Different Oxygen Percentages in CF$_4$/O$_2$ Plasma Generated by a High-Voltage Gas Discharge at a Discharge Current of 80 mA (Electrode Voltage $U = 0.8$ kV)

No.	Oxygen Percentage (%)	Average Etch Rate (nm/min)	Error (%)
1	0.5	48.2 ± 4.04	8.0
2	0.8	58.5 ± 2.34	4.0
3	8	22.5 ± 1.85	8.0
4	20	11.9 ± 1.04	9.0
5	35	10.3 ± 0.96	9.0
6	90	2.9 ± 0.23	8.0

TABLE A.29

Average Plasma-Chemical Etch Rates for Different Oxygen Percentages in CF_4/O_2 Plasma Generated by a High-Voltage Gas Discharge at a Discharge Current of 120 mA (Electrode Voltage $U = 0.8$ kV)

No.	Oxygen Percentage (%)	Average Etch Rate (nm/min)	Error (%)
1	0	17.8 ± 1.64	9.0
2	0.5	62.2 ± 3.43	6.0
3	0.8	72.5 ± 6.79	9.0
4	1.2	77.3 ± 5.86	8.0
5	8	34.8 ± 2.75	8.0
6	20	27.5 ± 2.70	10.0
7	50	12.6 ± 1.31	10.0
8	90	2.9 ± 0.23	8.0

TABLE A.30

Average Plasma-Chemical Etch Rates for Different Oxygen Percentages in CF_4/O_2 Plasma Generated by a High-Voltage Gas Discharge at a Discharge Current of 140 mA (Electrode Voltage $U = 0.8$ kV)

No.	Oxygen Percentage (%)	Average Etch Rate (nm/min)	Error (%)
1	0	10.6 ± 1.08	10.0
2	0.5	88.4 ± 9.21	10.0
3	0.8	94.5 ± 5.45	6.0
4	1.5	99.5 ± 10.17	10.0
5	3	84.5 ± 7.62	9.0
6	8	46.5 ± 4.14	9.0
7	50	18.3 ± 1.82	10.0
8	70	10.4 ± 0.90	9.0

TABLE A.31
Average Ion-Chemical Etch Rates for Different Oxygen Percentages in CF_4/O_2 Plasma Generated by a High-Voltage Gas Discharge at a Discharge Current of 50 mA (Electrode Voltage $U = 2$ kV)

No.	Oxygen Percentage (%)	Average Etch Rate (nm/min)	Error (%)
1	0	28.8 ± 2.15	7.0
2	0.2	34.8 ± 2.79	8.0
3	0.5	40.0 ± 4.18	10.0
4	1.2	39.4 ± 3.62	9.0
5	2	31.7 ± 2.70	9.0
6	4	25.5 ± 2.36	9.0
7	100	9.8 ± 0.94	10.0

TABLE A.32
Average Ion-Chemical Etch Rates for Different Oxygen Percentages in CF_4/O_2 Plasma Generated by a High-Voltage Gas Discharge at a Discharge Current of 80 mA (Electrode Voltage $U = 2$ kV)

No.	Oxygen Percentage (%)	Average Etch Rate (nm/min)	Error (%)
1	0	40.5 ± 2.74	7.0
2	0.5	212.7 ± 8.13	4.0
3	0.8	221.2 ± 7.90	4.0
4	1.2	214.9 ± 3.73	2.0
5	2	205.0 ± 5.59	3.0
6	10	83.3 ± 6.82	8.0
7	100	19.7 ± 2.05	10.0

TABLE A.33
Average Ion-Chemical Etch Rates for Different Oxygen Percentages in CF_4/O_2 Plasma Generated by a High-Voltage Gas Discharge at a Discharge Current of 120 mA (Electrode Voltage $U = 2$ kV)

No.	Oxygen Percentage (%)	Average Etch Rate (nm/min)	Error (%)
1	0	68.5 ± 4.78	7.0
2	0.2	199.7 ± 6.81	3.0
3	0.5	251.0 ± 6.32	3.0
4	0.8	260.0 ± 7.34	3.0
5	1.2	265.5 ± 6.62	2.0
6	10	119.5 ± 3.40	3.0
7	100	34.8 ± 2.91	8.0

TABLE A.34
Average Ion-Chemical Etch Rates for Different Oxygen Percentages in CF_4/O_2 Plasma Generated by a High-Voltage Gas Discharge at a Discharge Current of 140 mA (Electrode Voltage $U = 2$ kV)

No.	Oxygen Percentage (%)	Average Etch Rate (nm/min)	Error (%)
1	0	120.2 ± 5.71	5.0
2	0.5	220.7 ± 6.40	3.0
3	0.8	233.9 ± 4.90	2.0
4	1.2	240.1 ± 5.84	2.0
5	10	103.0 ± 5.11	5.0
6	100	21.1 ± 1.92	9.0

TABLE A.35
Average Plasma-Chemical Etch Rates for Different Substrate Temperatures in High-Voltage Gas-Discharge CF_4/O_2 Plasma at a Discharge Current of 50 mA

No.	Substrate Temperature (K)	Average Etch Rate (nm/min)	Error (%)
1	340	7.13 ± 0.419	6.0
2	360	9.9 ± 0.86	10.0
3	540	7.45 ± 0.639	9.0

TABLE A.36

Average Plasma-Chemical Etch Rates for Different Substrate Temperatures in High-Voltage Gas-Discharge CF_4/O_2 Plasma at a Discharge Current of 80 mA

No.	Substrate Temperature (K)	Average Etch Rate (nm/min)	Error (%)
1	340	52.7 ± 3.75	7.0
2	360	58.5 ± 2.34	4.0
3	540	41.9 ± 3.49	8.0

TABLE A.37

Average Plasma-Chemical Etch Rates for Different Substrate Temperatures in High-Voltage Gas-Discharge CF_4/O_2 Plasma at a Discharge Current of 120 Ma

No.	Substrate Temperature (K)	Average Etch Rate (nm/min)	Error (%)
1	340	73.5 ± 4.78	7.0
2	360	77.3 ± 5.86	8.0
3	540	63.5 ± 4.78	8.0

TABLE A.38

Average Plasma-Chemical Etch Rates for Different Substrate Temperatures in High-Voltage Gas-Discharge CF_4/O_2 Plasma at a Discharge Current of 140 mA

No.	Substrate Temperature (K)	Average Etch Rate (nm/min)	Error (%)
1	340	95.5 ± 8.33	9.0
2	360	99.5 ± 10.17	10.0
3	540	73.4 ± 7.17	10.0

TABLE A.39
Average Ion-Chemical Etch Rates for Different Substrate Temperatures in High-Voltage Gas-Discharge CF_4/O_2 Plasma at a Discharge Current of 50 mA

No.	Substrate Temperature (K)	Average Etch Rate (nm/min)	Error (%)
1	320	25.2 ± 1.25	5.0
2	360	32.1 ± 2.64	8.0
3	395	22.2 ± 1.64	7.0

TABLE A.40
Average Ion-Chemical Etch Rates for Different Substrate Temperatures in High-Voltage Gas-Discharge CF_4/O_2 Plasma at a Discharge Current of 80 mA

No.	Substrate Temperature (K)	Average Etch Rate (nm/min)	Error (%)
1	370	48.3 ± 3.07	6.0
2	390	221.2 ± 7.90	4.0
3	440	24.0 ± 2.08	9.0

TABLE A.41
Average Ion-Chemical Etch Rates for Different Substrate Temperatures in High-Voltage Gas-Discharge CF_4/O_2 Plasma at a Discharge Current of 120 mA

No.	Substrate Temperature (K)	Average Etch Rate (nm/min)	Error (%)
1	380	24.4 ± 2.36	10.0
2	390	39.5 ± 2.36	6.0
3	422	265.5 ± 6.62	2.0
4	470	40.0 ± 2.67	7.0

TABLE A.42

Average Ion-Chemical Etch Rates for Different Substrate Temperatures in High-Voltage Gas-Discharge CF_4/O_2 Plasma at a Discharge Current of 140 mA

No.	Substrate Temperature (K)	Average Etch Rate (nm/min)	Error (%)
1	375	20.1 ± 1.95	10.0
2	390	39.9 ± 4.17	10.0
3	410	80.0 ± 5.33	7.0
4	440	240.1 ± 5.84	2.0
5	480	45.2 ± 4.01	9.0

References

1. Soifer, V. A., ed. 2014. *Diffractive Nanophotonics*. London: Taylor & Francis.
2. Soifer, V. A., ed. 2012. *Computer Design of Diffractive Optics*. Cambridge: Woodhead Publishing and Cambridge International Science Publishing.
3. Sundaramurthy, A., Schuck, P. J., Conley, N. R., Fromm, D. P., Kino, G. S., and Moerner, W. E. 2006. Toward nanometer-scale optical photolithography: Utilizing the near-field of Bowtie optical nanoantennas. *Nano Letters* 6(3), 355–360.
4. Neisser, M. and Wurm, S. 2015. ITRS lithography roadmap: 2015 challenges. *Advanced Optical Technologies* 4(4), 235–240.
5. Denisyuk, Yu, N., Ganzherli, N. M., and Chernykh, D. F. 2000. Multiplexed counter-beam speckle holograms in bichromated gelatin. *Technical Physics Letters* 26(5), 369–371.
6. Smirnova, T. N., Sarbaev, T. A., and Tikhonov, E. A. 1994. Real-time holographic recording of reflection gratings in a photopolymerisable composite. *Quantum Electronics* 24(4), 348–350.
7. Volkov, A. V., Kazanskiy, N. L., Moiseyev, O. Yu., and Soifer, V. A. 1998. A method for the diffractive microrelief formation using the layered photoresist growth. *Optics and Lasers in Engineering* 29(4–5), 281–288.
8. Yang, G., ed. 2012. *Laser Ablation in Liquids: Principles and Applications in the Preparation of Nanomaterials*. New York: Taylor & Francis.
9. Liu, Q., Duan, X., and Peng, C. 2014. *Novel Optical Technologies for Nanofabrication*. Berlin: Springer Berlin Heidelberg.
10. Gaol, W., Singh, N., Song, L., Liu, Z., Reddy, A. L. M., Ci, L., Vajtai, R., Zhang, Q., Wei, B., and Ajayan, P. M. 2011. Direct laser writing of micro-supercapacitors on hydrated graphite oxide films. *Nature Nanotechnology* 6, 496–500.
11. Sainiemi, L., Jokinen, V., Shah, A., Shpak, M., Aura, S., Suvanto, P., and Franssila, S. 2011. Non-reflecting silicon and polymer surfaces by plasma etching and replication. *Advanced Materials* 23(1), 122–126.
12. Oehrlein, G. S., Phaneuf, R. J., and Graves, D. B. 2011. Plasma-polymer interactions: A review of progress in understanding polymer resist mask durability during plasma etching for nanoscale fabrication. *Journal of Vacuum Science and Technology B* 29(1), 010801-1–010801-35.
13. Korolkov, V. P., Ostapenko, S. V., Nasyrov, R. K., Gutman, A. S., and Sametov, A. R. 2010. Unification of approaches to optimization and metrological characterization of continuous-relief diffractive optical elements. In *Optical Micro- and Nanometrology III, Proc. SPIE, 7718*, Edited by Gorecki, C., Asundi, A. K. and Osten, W. Brussels: SPIE. p. 77180S-1–77180S-9.
14. Xu, H., Lu, N., Shi, G., Qi, D., Yang, B., Li, H., Xu, W., and Chi, L. 2011. Biomimetic antireflective hierarchical arrays. *Langmuir* 27(8), 4963–4967.
15. Mekaru, H., Koizumi, O., Ueno, A., and Takahashi, M. 2010. Inclination of mold pattern's sidewalls by combined technique with photolithography at defocus-positions and electroforming. *Microsystem Technologies* 16(8), 1323–1330.
16. Volkov, A. V. 2002. Methods and Experimental Setups for Fabricating the Microreliefs of Visible and Infrared Diffractive Optical Elements. D. Tech. Sc. dissertation, Presented March 7, 2002, approved September 26, 2002. Samara (in Russian).
17. Moiseyev, O. Yu. 2000. Research into Methods for Fabricating Infrared Diffractive Optical Elements by Using Photoresists and Photopolymer Compositions. Cand. Tech. Sc. dissertation, Presented June 7, 2000, approved December 26, 2000. Samara (in Russian).

18. Reinhardt, K. A. and Kern, W. 2008. *Handbook of Silicon Wafer Cleaning Technology.* Norwich: William Andrew.

19. Sparks, C. M. 2011. Novel analytical methods for cleaning evaluation: Chapter 15. In *Handbook of Cleaning for Semiconductor Manufacturing,* Edited by Reinhardt, K. A. and Rcidy, R. p. 543–564. Norwich: William Andrew.

20. Borodin, S. A., Volkov, A. V., and Kazanskiy, N. L. 2009. Device for analyzing nano-roughness and contamination on a substrate from the dynamic state of a liquid drop deposited on its surface. *Journal of Optical Technology* 76(7), 408–412.

21. Czanderna, A. W., ed. 1989. *Methods of Surface Analysis.* New York: Elsevier Science Publishers.

22. Woodruff, D. P. and Delchar, T. A. 2003. *Modern Techniques of Surface Science.* Cambridge: Cambridge University Press.

23. Shemesh, D., Eilon, M., Doozli, Hen., Rechav, E., and Binyamini, E. February 9, 2006, filed, and January 29, 2013, published. Method for monitoring a surface's cleanliness. US Patent 8361814.

24. Gorin, A., Jaouad, A., Grondin, E., Aimez, V., and Charette, P. 2008. Fabrication of silicon nitride waveguides for visible-light using PECVD: A study of the effect of plasma frequency on optical properties. *Optics Express* 16(18), 13509–13516.

25. Bradley, J. D. B., Ay, F., Worhoff, K., and Pollnau, M. 2007. Fabrication of low-loss channel waveguides in Al_2O_3 and Y_2O_3 layers by inductively coupled plasma reactive ion etching. *Applied Physics B* 89(2), 311–318.

26. Lee, C. L., Gu, E., Dawson, M. D., Friel, I., and Scarsbrook, G. A. 2008. Etching and micro-optics fabrication in diamond using chlorine-based inductively-coupled plasma. *Diamond and Related Materials* 17(7–10), 1292–1296.

27. Yamamura, K. 2007. Fabrication of ultra precision optics by numerically controlled local wet etching. *CIRP Annals—Manufacturing Technology* 56(1), 541–544.

28. Karlsson, M., Hjort, K., and Nikolajeff, F. 2001. Transfer of continuous-relief diffractive structures into diamond by use of inductively coupled plasma dry etching. *Optics Letters* 26(22), 1752–1754.

29. Ruan, Y., Li, W., Jarvis, R., Madsen, N., Rode, A., and Barry Luther-Davies, B. 2004. Fabrication and characterization of low loss rib chalcogenide waveguides made by dry etching. *Optics Express* 12(21), 5140–5145.

30. Fu, Y., Kok, B., and Ngoi, A. 2001. Investigation of diffractive-refractive micro-lens array fabricated by focused ion beam technology. *Optical Engineering* 40(4), 511–516.

31. Ekberg, M. et al. 2001. Laser-triggered high-voltage plasma switching with diffractive optics. *Applied Optics* 40(16), 2611–2617.

32. Li, C. and Nikumb, S. 2003. Optical quality micromachining of glass with focused laser-produced metal plasma etching in the atmosphere. *Applied Optics* 42(13), 2383–2387.

33. Takino, H. et al. 2002. Fabrication of optics by use of plasma chemical vaporisation machining with a pipe electrode. *Applied Optics* 41(19), 3971–3977.

34. Shul, R. J. and Pearton, S. J. 2000. *Handbook of Advanced Plasma Processing Techniques.* Berlin: Springer Berlin Heidelberg.

35. Shamiryan, D., Paraschiv, V., Boullart, W., and Baklanov, M. R. 2009. Plasma etching: From micro- to nanoelectronics. *High Energy Chemistry* 43(3), 204–212.

36. Pierson, J. F., Wiederkehr, D., and Billard, A. 2005. Reactive magnetron sputtering of copper, silver, and gold. *Thin Solid Films* 478(1–2), 196–205.

37. Schoenbach, K. H., Verhappen, R., Tessnow, T., Peterkin, F. E., and Byszewski, W. W. 1996. Microhollow cathode discharges. *Applied Physics Letters* 68(1), 13–15.

38. Stark, R. H. and Schoenbach, K. H. 1999. Direct current high-pressure glow discharges. *Applied Physics* 85(4), 2075–2080.

39. Gavrilov, N. V., Emlin, D. R., and Nikulin, S. P. 1999. Generation of a homogeneous plasma in a glow discharge with a hollow anode and a wide-aperture hollow cathode. *Technical Physics Letters* 25(6), 498–500.

40. Nikulin, S. P. and Kuleshov, S. V. 2000. Generation of homogeneous plasma in a low-pressure glow discharge. *Technical Physics* 45(4), 400–405.

41. Šišović, N. M., Majstorović, G. Lj., and Konjević, N. 2007. Excessive Doppler broadening of the Hα line in a hollow cathode glow discharge. *The European Physical Journal D* 41(1), 143–150.

42. Pinnaduwage, L. A., Ding, W., and McCorkle, D. L. 1997. Enhanced electron attachment to highly excited molecules using a plasma mixing scheme. *Applied Physics Letters* 71(25), 3634–3636.

43. Nakagawa, S., Tahara, Y., and Ogasawara, M. October 20, 1994, filed, and September 12, 1995, published. Apparatus and method for generating plasma of uniform flux density. US Patent 5449977.

44. Yasaka, Y. and Nakamura, T. 1996. Control of process uniformity by using electron–cyclotron resonance plasma produced by multiannular antenna. *Applied Physics Letters* 68(11), 1476–1478.

45. Foster, D. F. and Wu, Ch.-H. J. 1996. Assessment of self-consistent analytic model for inductive RF discharge and design of uniform discharge with planar-vertical antennas. *IEEE Transactions on Plasma Sciences* 24(3), 1155–1163.

46. Sung-Spitzl, H. June 23, 1997, filed, and January 28, 1999, published. Vorrichtung zur Erzeugung von homogenen Mikrowellenplasmen [Device for the production of homogenous microwave plasma]. Patent Application DE 19726663.

47. Chabert, P. and Braithwaite, N. 2011. *Physics of Radio-Frequency Plasmas*. Cambridge: Cambridge University Press.

48. Sittsworth, J. A. and Wendt, A. E. 1996. Reactor geometry and plasma uniformity in a planar inductively coupled radio frequency argon discharge. *Plasma Sources Science and Technology* 5(3), 429–435.

49. Uedo, Y., Muta, H., and Kawai, Y. 1999. Role of peripheral vacuum regions in the control of the electron–cyclotron resonance plasma uniformity. *Applied Physics Letters* 74(14), 1972–1974.

50. Korzec, D., Werner, F., Winter, R., and Engemann, J. 1996. Scaling of microwave slot antenna (SLAN): A concept for efficient plasma generation. *Plasma Sources Science and Technology* 5(2), 216–234.

51. Stittsworth, J. A. and Wendt, A. E. 1996. Striations in a radio frequency planar inductively coupled plasma. *IEEE Transactions on Plasma Sciences* 24(1), 125–126.

52. Doh, H.-H., Yeon, Ch.-K., and Whang, K.-W. 1997. Effects of bias frequency on reactive ion etching lag in an electron–cyclotron resonance plasma etching system. *Journal of Vacuum Science and Technology A* 15(3-1), 664–667.

53. Kovalevskii, A. A., Malyshev, V. S., Tsybul'skii, V. V., and Sorokin, V. M. 2002. Isotropic plasma etching of SiO₂ films. *Russian Microelectronics* 31(5), 290–294.

54. Putrya, M. G. 2005. *Plasma Methods for Generating Three-Dimensional ULIC Patterns*. Moscow: MIET Publisher (in Russian).

55. Woodworth, J. R., Aragon, B. P., and Hamilton, T. W. 1997. Effect of bumps on the wafer on ion distribution functions in high-density argon and argon-chlorine discharges. *Applied Physics Letters* 70(15), 1947–1949.

56. Hebner, G. A., Blain, M. G., and Hamilton, T. W. 1999. Influence of surface material on the boron chloride density in inductively coupled discharges. *Journal of Vacuum Science and Technology A* 17(6), 3218–3224.

57. Miyata, K., Hori, M., and Goto, T. 1996. CF$_x$ radical generation by plasma interaction with fluorocarbon films on the reactor wall. *Journal of Vacuum Science and Technology A* 14(4), 2083–2087.

58. Komine, K., Araki, N., Noge, S., Ueno, H., and Hohkawa, K. 1996. Residuals caused by the CF_4 gas plasma etching process. *Japanese Journal of Applied Physics* 35(5b-1), 3010–3014.

59. McLane, G. F., Dubey, M., Wood, M. C., and Lynch, K. E. 1997. Dry etching of germanium in magnetron enhanced SF_6 plasmas. *Journal of Vacuum Science and Technology B* 15(4), 990–992.

60. Stoffels, W. W., Stoffels, E., and Tachibana, K. 1998. Polymerisation of fluorocarbons in reactive ion etching plasmas. *Journal of Vacuum Science and Technology A* 16(1), 87–95.

61. Schwarzenbach, W., Cunge, G., and Booth, J. P. 1999. High mass positive ions and molecules in capacitively coupled radio-frequency CF_4 plasmas. *Journal of Applied Physics* 85(11), 7562–7568.

62. Mieno, T. and Samukawa, S. 1997. Generation and extinction characteristics of negative ions in pulse-time-modulated electron–cyclotron resonance chlorine plasma. *Plasma Sources Science and Technology* 6(3), 398–404.

63. Vagner, I. V., Bolgov, E. I., Grakun, V. F., Gokhveld, V. L., and Kudlai, V. A. 1974. Elementary cell for generating arbitrarily shaped electron beams under high-voltage discharge in gas. *Zhurnal tekhnicheskoi fiziki* 44(8), 1669–1674 (in Russian).

64. Komov, A. N., Kolpakov, A. I., Bondareva, N. I., and Zakharenko, V. V. 1984. Electron-beam setup for welding the components of semiconductor devices. *Pribory i tekhnika eksperimenta* 5, 218–220 (in Russian).

65. Markushin, M. A., Kolpakov, V. A., Krichevskii, S. V., and Kolpakov, A. I. 2015. Simulation of the electric field distribution in the electrode system of a device forming a high-voltage gas discharge. *Technical Physics* 60(3), 376–380.

66. Kolpakov, V. A., Kolpakov, A. I., and Kritchevsky, S. V. 1996. Ion-plasma cleaning of low-power relay contacts. *Electronics Industry* 2, 41–44.

67. Vagner, I. V., Bolgov, E. I., Grakun, V. F., Gokhveld, V. L., and Kudlai, V. A. 1972. Generation of arbitrarily shaped electron beams under high-voltage discharge in gas and specific features of designing the gas-discharge guns. *Avtomaticheskaya svarka* 12, 27 (in Russian).

68. Handle, S. K. and Nordhage, F. R. 1997. Method for triggering high-voltage vacuum discharges. *Journal of Applied Physics* 81(9), 6473–6475.

69. Donko, Z., Rozsa, K., and Szalai, L. 1998. High-voltage hollow cathode discharges: Laser applications and simulations of electron motion. *Plasma Physics Reports* 24(7), 588.

70. Kolpakov, V. A., Kolpakov, A. I., and Kritchevsky, S. V. 1995. A device for quickly measuring the surface cleanliness of dielectric substrates. *Instruments and Experimental Techniques* 5, 199–200.

71. Kazanskiy, N. L., Kolpakov, V. A., Kolpakov, A. I., Kritchevsky, S. V., and Ivliev, N. A. 2005. Parameter optimisation of a tribometric device for rapid assessment of substrate surface cleanliness. *Computer Optics* 28, 76–79.

72. Orlikovskiy, A. A. and Rudenko, K. V. 2001. *In situ* diagnostics of plasma processes in microelectronics: The current status and immediate prospects. Part I. *Russian Microelectronics* 30(2), 69–87.

73. Orlikovskiy, A. A., Rudenko, K. V., and Sukhanov, Ya. N. 2001. *In situ* diagnostics of plasma processes in microelectronics: The current status and immediate prospects. Part II. *Russian Microelectronics* 30(3), 163–182.

74. Orlikovskiy, A. A. and Rudenko, K. V. 2001. *In situ* diagnostics of plasma processes in microelectronics: The current status and immediate prospects. Part III. *Russian Microelectronics* 30(5), 275–294.

75. Orlikovskiy, A. A., Rudenko, K. V., and Sukhanov, Ya. N. 2001. *In situ* diagnostics of plasma processes in microelectronics: The current status and immediate prospects. Part IV. *Russian Microelectronics* 30(6), 343–370.

76. Averkin, S. N., Yershov, A. P., Orlikovskiy, A. A., Rudenko, K. V., and Sukhanov, Ya. N. 2003. Comparative study of an RF and a microwave high-density-plasma source for plasma immersion ion implantation. *Russian Microelectronics* 32(5), 292–300.
77. Orlikovskiy, A. A., Rudenko, K. V., and Averkin, S. N. 2006. Fine-line plasma-enhanced processes on the basis of a set of pilot units with a scalable inductively coupled plasma source for use in microelectronics. *High Energy Chemistry* 40(3), 182–193.
78. Chu, P. K. and Lu, X. P., eds. 2010. *Low Temperature Plasma Technology: Methods and Applications*. London: Taylor & Francis.
79. Fadeev, A. V., Rudenko, K. V., Lukichev, V. F., and Orlikovskii, A. A. 2011. Optimization of the tomographic algorithm of the reconstruction of plasma irregularities in process reactors of microelectronics. *Russian Microelectronics* 40(2), 108–118.
80. Denisova, N. 2009. Plasma diagnostics using computed tomography method. *IEEE Transactions on Plasma Science* 37(4), 502–512.
81. Artsimovich, L. A. and Lukyanov, S. Yu. 1972. *Motion of Charged Particles in Electric and Magnetic Fields*. Moscow: Nauka Publisher (in Russian).
82. Magunov, A. N. 1999. Heat-transfer instabilities in the interaction of a nonequilibrium plasma with a solid surface. *Plasma Physics Reports* 25(8), 646–651.
83. Magunov, A. N. 2000. Laser thermometry of solids in plasma (Review). *Instruments and Experimental Techniques* 43(2), 149–172.
84. Kolpakov, V. A., Kolpakov, A. I., and Podlipnov, V. V. 2013. Formation of an out-of-electrode plasma in a high-voltage gas discharge. *Technical Physics* 58(4), 505–510.
85. Vagner, I. V., Grakun, V. F., and Gokhveld, V. L. 1971. Forming electron and ion beams in high-voltage gas discharge. *Avtomaticheskaya svarka* 12, 15–17 (in Russian).
86. Kolpakov, V. A. and Kolpakov, A. I. 2001. Plasma-chemical etching of silicon dioxide in high-voltage gas-discharge plasma. In *Proceedings of BIKAMP-2001, The Third International Conference on the Future of Information, Space, Aircraft, and Medicine Instrumentation*. p. 90–92. Saint Petersburg (in Russian).
87. Rykalin, N. N., Zuev, I. V., and Uglov, I. V. 1978. *Principles of Electron-Beam Treatment of Materials*. Moscow: "Mashinostroenie" Publisher (in Russian).
88. Kazanskiy, N. L. and Kolpakov, V. A. 2003. Study of principles of generating low-temperature plasma by high-voltage gas discharge. *Computer Optics* 25, 112–117 (in Russian).
89. Molokovsky, S. I. and Sushkov, A. D. 1991. *High-Intensity Electron and Ion Beams*. Moscow: Energoatomizdat Publisher (in Russian).
90. Soifer, V. A., Kazanskiy, N. L., Kolpakov, V. A., Kolpakov, A. I., Paranin, V. D., and Desyatov, M. V. March 11, 2008, filed, and September 10, 2009, published. Method for determination of charged particles flow. Patent RU 2366978 (in Russian).
91. Kolpakov, A. I. and Kolpakov, V. A. 1999. Dragging of silicon atoms by vacancies created in molten aluminium under ion–electron irradiation. *Pisma v ZhTF* 25(15), 618.
92. Chernetsky, A. V. 1969. *Introduction to Plasma Physics*. Moscow: Atomizdat Publisher (in Russian).
93. Kazanskiy, N. L. and Kolpakov, V. A. 2006. Studies into mechanisms of generating a low-temperature plasma in high-voltage gas discharge. *Optical Memory and Neural Networks* 15(4), 163–169.
94. Raizer, Yu. P. 1991. *Gas Discharge Physics*. Berlin: Springer Berlin Heidelberg.
95. Chernyaev, V. N. 1987. *Physical and Chemical Processes in Electronics Manufacture*. Moscow: Vysshaya shkola Publisher (in Russian).
96. Izmailov, S. V. 1939. On the thermal theory of electron emission under the impact of fast ions. *Zhurnal eksperimentalnoi i teoreticheskoi fiziki* 9(12), 1473–1483 (in Russian).
97. Kazanskiy, N. L., Kolpakov, V. A., Kritchevsky, S. V., and Paranin, V. D. 2006. Investigation into formation mechanisms of high-voltage gas-discharge plasma.

In *Proceedings of an All-Russian Scientific and Technical Conference Titled Methods for Developing and Researching Materials and Equipment and Economic Aspects of Microelectronics.* p. 35–39. Penza (in Russian).

98. Mataré, H. F. 1974. *Elektronika defektov v poluprovodnikakh.* Translated by Guro, G. M., Edited by Medvedev, S. A. Moscow: "Mir" Publisher. Originally published as 1971. *Defect Electronics in Semiconductors.* New York: John Wiley and Sons.

99. Rozanov, L. N. 2002. *Vacuum Technique.* London: Taylor & Francis.

100. Kazanskiy, N. L., Kolpakov, V. A., Kolpakov, A. I., and Kritchevsky, S. V. June 14, 2005, filed, and March 20, 2007, published. Cable for supplying low-temperature plasma generator with power. Patent RU 2295791 (in Russian).

101. Kolpakov, V. A., Paranin, V. D., and Mokeyev, D. A. 2007. Cable for supplying low-temperature plasma generator with power. In *Proceedings of an International Scientific Workshop Titled Modern Equipment and Technologies.* 1, p. 86–88. Tomsk (in Russian).

102. Oks, E., Vizir, A., and Yushkov, G. 1998. Development of vacuum arc ion sources for heavy ion accelerator injectors and ion implantation technology. *Review of Scientific Instruments* 69(2), 853.

103. Knechtli, R. May 25, 1973, filed, and August 20, 1974, published. Hollow cathode gas-discharge device. US Patent 3831052 A.

104. Burdovitsin, V. A., Fedorov, M. V., and Oks, E. M. March 24, 2003, filed, and June 20, 2004, published. Plasma electron source of the ribbon beam. Patent RU 2231164 (in Russian).

105. Burdovitsin, V. A., Kuzemchenko, M. N., and Oks, E. M. February 8, 2002, filed, and October 27, 2003, published. Plasma electron source. Patent RU 2215383 (in Russian).

106. Kazanskiy, N. L., Kolpakov, V. A., and Podlipnov, V. V. 2014. Gas discharge devices generating the directed fluxes of off-electrode plasma. *Vacuum* 101, 291–297.

107. Soifer, V. A., Kazanskiy, N. L., Kolpakov, V. A., Kolpakov, A. I., and Podlipnov, V. V. December 25, 2006, filed, and November 20, 2008, published. Focuser of gas-discharge plasma. Patent RU 2339191 (in Russian).

108. Moreau, W. 1990. Mikrolitografiya: Printsipy, metody, materialy. Part 1. Edited by Timerov, R. Kh. Moscow: Mir Publisher. Originally published as (1988) *Semiconductor Lithography: Principles, Practices and Materials.* New York: Plenum Press.

109. Poltavtzev, Yu. G., and Knyazev, A. S. 1990. *Techniques of Surface Processing in Microelectronics.* Kiev: Tekhnika Publisher (in Russian).

110. Oura, K., Lifshits, V. G., Saranin, A. A., Zotov, A. V., and Katayama, M. 2003. *Surface Science: An Introduction.* Berlin: Springer Berlin Heidelberg.

111. Westphal, O. L. and Myagkov, A. T. 1976. Methods and tools for monitoring the cleanliness and quality of semiconductor surfaces. *Obzory po elektronnoi tekhnike.* 6th ser., *Materialy* 9 (in Russian).

112. Bogatyrev, A. Ye., Shushunova, L. I., and Tsyganov, G. M. 1980. New methods for monitoring surface cleanliness and defects of parts. *Obzory po elektronnoi tekhnike* 3, 19–27 (in Russian).

113. Harrick, N. 1970. *Spektroskopiya vnutrennego otrazheniya.* Translated by Zolotarev, V. M. and Berstein, V. A. Moscow: Mir Publisher. Originally published as *Internal Reflection Spectroscopy.* 1967. New York: Interscience Publishers (in Russian).

114. Zharkikh, Yu. S., Pastushenko, A. M., Misyura, A. V., and Tronko, T. V. 1977. The effect of chemical treatments on the heterogeneity of silicon surface potential. *Poluprovodnikovaya tekhnika i mikroelektronika.* 25, 40–44 (in Russian).

115. Volkenstein, F. F. 1987. *Electronic Processes on Semiconductor Surfaces during Chemisorption.* Moscow: Nauka Publisher (in Russian).

116. Zharkikh, Yu. S., Yevdokimov, A. D., Poltavtsev, Yu. G., and Levitskaya, R. O. 1983. Comparison of two monitoring methods for pre-oxidation treatments of silicon wafers. *Optoelektronika i poluprovodnikovaya tekhnika* 4, 3–4 (in Russian).
117. Kolpakov, A. I. 1993. Method for determining the surface cleanliness of substrates. *Elektronnaya promishlennost* 4, 37–39 (in Russian).
118. Borodin, S. A., Volkov, A. V., Kolpakov, A. I., and Rafelson, L. L. 1990. A device for monitoring the surface cleanliness of substrates. *Pribory i tekhnika eksperimenta* 5, 230–232 (in Russian).
119. Frolov, Ye. S., Minaychev, Ye. V., and Aleksandrova, A. T. 1985. *Vacuum Engineering: A Handbook*. Edited by Frolov, Ye. S. Moscow: Mashinostroyenie Publisher (in Russian).
120. Hebda, M. 1989. A handbook of triboengineering. In *Theoretical Foundations*, Vol. 1. Edited by Hebda, M. and Chichinadze, A. V. Moscow: Mashinostroyenie Publisher (in Russian).
121. Kragelsky, I. V., Dobychin, M. N., and Kombalov, V. S. 1982. *Friction and Wear: Calculation Methods*. Oxford: Pergamon Press.
122. Mikhin, N. M. 1993. External friction of solids. In *Tribology: Research and Application; Experience in USA and CIS Countries*. Edited by Beliy, V. A., Ludema, K. and Myshkin, N. K. p. 29–51. Moscow: Mashinostroyenie Publisher; New York: Allerton Press.
123. Volkov, A. V. and Kolpakov, A. I. April 2, 1990, filed, and October 12, 1992, published. Method for measuring the cleanliness of substrate surfaces. USSR author's certificate 1821688 (in Russian).
124. Kazanskiy, N. L., Kolpakov, V. A., Kritchevsky, S. V., and Ivliev, N. A. June 14, 2005, filed, and September 27, 2007, published. Method for measuring the cleanliness of substrate surfaces. Patent RU 2307339 (in Russian).
125. Grechishnikov, V. M. 2007. *Metrology and Radio Measurements: A Textbook*. Samara: Samara State Aerospace University.
126. Kazanskiy, N. L., Kolpakov, V. A., Kolpakov, A. I., Kritchevsky, S. V., and Ivliev, N. A. 2007. Interaction of dielectric substrates in the course of tribometric assessment of the surface cleanliness. *Computer Optics* 31(1), 42–46.
127. Kazanskiy, N. L., Kolpakov, V. A., Kolpakov, A. I., Kritchevsky, S. V., and Desyatov, M. V. 2008. Parameter optimisation of a tribometric device for rapid assessment of substrate surface cleanliness. *Optical Memory and Neural Networks* 17(2), 167–172.
128. Danilin, B. S. and Kireev, V. Yu. 1987. *Application of Low-Temperature Plasma for Etching and Cleaning of Materials*. Moscow: Energiya Publisher (in Russian).
129. Buckley, D. H. 1989. Poverkhnostniye yavleniya pri adgezii i friktsionnom vzaimodeystvii. Translated by Belyi, A. V. and Myshkin, N. K. Edited by Sviridenko, A. I. Moscow: Mashinostroyenie Publisher. Originally published in 1981. *Surface Effects in Adhesion, Friction, Wear and Lubrication*. Amsterdam: Elsevier Science.
130. Kazanskiy, N. L., Kolpakov, V. A., Karpeev, S. V., Kritchevsky, S. V., and Ivliev, N. A. 2008. Interaction of dielectric substrates in the course of tribometric assessment of the surface cleanliness. *Optical Memory and Neural Networks* 17(1), 37–42.
131. Kazanskiy, N. L., Kolpakov, V. A., Kolpakov, A. I., Ivliev, N. A., and Kritchevsky, S. V. 2013. Device for checking the surface finish of substrates by tribometry method. *Friction and Wear Research* 1(1), 10–14.
132. Borodin, S. A., Volkov, A. V., and Kazanskiy, N. L. 2009. Device for analysing nanoroughness and contamination on a substrate from the dynamic state of a liquid drop deposited on its surface. *Journal of Optical Technology* 76(7), 408–412.
133. Glagolev, K. V. and Morozov, A. N. 2002. *Physical Thermodynamics*. Chapter 7. Moscow: Bauman Moscow State Technical University (in Russian).

134. Pastukhov, V. A., Bokser, E. L., and Tsarevskiy, B. V. 1976, July 30. A method for determining the contact angle of solids. USSR author's certificate N531065 (in Russian).

135. Itsiksonas, G. O. and Schumacher, A. A. 1977, February 17. A method for determining a contact angle. USSR author's certificate N548788 (in Russian).

136. Magunov, A. N. 1986. September 30. A method for determining a contact angle. USSR author's certificate N1260752 (in Russian).

137. Magunov, A. N. and Mudrov, Ye. V. 1990. Measuring the contact angle with the reflected-light method. *Pribory i tekhnika eksperimenta* 5, 227–230 (in Russian).

138. Zisman, W. 1964. *Contact Angle, Wettability and Adhesion*, Fowkes, F. M., ed., Washington: American Chemical Society.

139. Padday, J. F., ed. 1978. *Wetting, Spreading and Adhesion*, New York: Academic.

140. de Gennes, P. G. 1985. Wetting: Statics and dynamics. *Reviews of Modern Physics* 57(3), 827–863.

141. Rafelson, L. L., Volkov, A. V., Borodin, S. A., and Ivanova, V. A. 1992, February 15. A device for monitoring the cleanliness of substrate surfaces. USSR author's certificate 1741032 (in Russian).

142. Volkov, A. V. and Kolpakov, A. I. 1992, September 1. A method for determining the cleanliness of substrate surfaces. USSR author's certificate 1784868 (in Russian).

143. Kazanskiy, N. L., Volkov, A. V., and Borodin, S. A. July 17, 2006, filed, and August 20, 2008, published. Method of surface finish control for dielectric substrates. Patent RU 2331870 (in Russian).

144. Millman, J. and Grabel, A. 1987. *Microelectronics*, 2nd ed. New York: McGraw-Hill, Inc.

145. Izotov, P. Yu., Glyanko, M. S., Volkov, A. V., Kazanskiy, N. L., and Sukhanov, S. V. November 29, 2010, filed, and April 20, 2012, published. Device to control roughness of dielectric substrate surfaces. Patent RU 2448341 (in Russian).

146. Izotov, P. Yu., Glyanko, M. S., and Sukhanov, S. V. 2011. Modification of device for displaying the finish and smoothness of optical substrates. *Computer Optics* 35(1), 63–69.

147. Glyanko, M. S. and Izotov, P. Yu. 2012. Software for the device for detection of cleanliness and roughness of optical substrates. *Computer Optics* 36(2), 42–248.

148. Izotov, P. Yu. 2013. Integration of an automated dispenser into the surface cleanliness and roughness analyser for optical substrates. *Izvestiya Samarskogo nauchnogo tsentra RAN* 15(4-1), 154–158 (in Russian).

149. Golub, M. A., Kazanskiy, N. L., Sisakyan, I. N., and Soifer, V. A. 1990. Wavefronts forming by computer-generated optical elements. *Proceedings of SPIE* 1183, 727–750.

150. Kazanskiy, N. L., Kharitonov, S. I., and Soifer, V. A. 2000. Simulation of DOE-aided focusing devices. *Optical Memory and Neural Networks* 9(3), 191–200.

151. Kazanskiy, N. L., Kharitonov, S. I., Soifer, V. A., and Volkov, A. V. 2000. Investigation of lighting devices based on diffractive optical elements. *Optical Memory and Neural Networks* 9(4), 301–312.

152. Moiseyev, M. A., Doskolovich, L. L., and Kazanskiy, N. L. 2011. Design of high-efficient freeform LED lens for illumination of elongated rectangular regions. *Optics Express* 19(S3), A225–A233.

153. Doskolovich, L. L., Kazanskiy, N. L., Soifer, V. A., Kharitonov, S. I., and Perlo, P. 2004. A DOE to form a line-shaped directivity diagram. *Journal of Modern Optics* 51(13), 1999–2005.

154. Doskolovich, L. L., Kazanskiy, N. L., Kharitonov, S. I., Perlo, P., and Bernard, S. 2005. Designing reflectors to generate a line-shaped directivity diagram. *Journal of Modern Optics* 52(11), 1529–1536.

155. Doskolovich, L. L., Kazanskiy, N. L., and Bernard, S. 2007. Designing a mirror to form a line-shaped directivity diagram. *Journal of Modern Optics* 54(3–4), 589–597.

156. Volotovskii, S. G., Kazanskiy, N. L., Popov, S. B., and Khmelev, R. V. 2005. Machine vision system for registration of oil tank wagons. *Pattern Recognition and Image Analysis* 15(2), 461–463.

157. Kazanskiy, N. L. and Popov, S. B. 2010. Machine vision system for singularity detection in monitoring the long process. *Optical Memory and Neural Networks (Information Optics)* 19(1), 23–30.

158. Hamilton Company. 2013. Microlab 600 standalone pump: Ordering information. Available from: http://www.hamiltoncompany.com/products/microlab-600/c/1170/ (webpage discontinued). (Accessed April 12, 2013).

159. Google. 2013. Giesecke, R. C# Project Template for Unmanaged Exports. Available from: https://sites.google.com/site/robertgiesecke/Home/uploads/. (Accessed April 12, 2013).

160. Duval, P. 1988. *High Vacuum Production in the Microelectronics Industry*. Amsterdam: Elsevier.

161. Anishchenko, E. V., Affiliated with OAO Nauchno-issledovatel'skii institut poluprovodnikovykh priborov Kagadei, V. A., Nefedtsev, E. V., Oskomov, K. V., Proskurovski, D. I., and Romanenko, S. V. 2005. Residual-photoresist removal from Si and GaAs surfaces by atomic-hydrogen flow treatment. *Russian Microelectronics* 34(3), 131–139.

162. Nefedov, D. V. and Yafarov, R. K. 2004. The effect of microwave radiation–plasma interaction on the process parameters of plasma treatment. In *Proceedings of a Scientific and Technical Conference Titled Current Problems of Electronic Engineering*. p. 334–339. Saratov (in Russian).

163. Vishnyakov, A. S., Kagadei, V. A., Kozhinova, N. I., Proskurovsky, D. I., and Romas, L. M. 2000. Surface cleaning with an atomic hydrogen flow in fabricating ohmic and barrier contacts to GaAs and AlxGa1-xAs. *Russian Microelectronics* 28(6), 442–453.

164. Piganov, M. N., Kritchevsky, S. V., and Kolpakov, V. A. 1997. Improving the reliability and quality of nonregularly structured microassemblies. In *Proceedings of a Scientific Conference Titled Actual Problems of Analysing and Ensuring the Reliability and Quality of Equipment, Facilities, and Systems*. Edited by Tartakovskiy, A. M. and Bloinov, A. V. p. 32–33. Penza: Penza University Publisher (in Russian).

165. Kazanskiy, N. L., Kolpakov, V. A., and Kritchevsky, S. V. 2006. Simulating the process of dielectric-substrate surface cleaning in high-voltage gas-discharge plasma. *Proceedings of SPIE* 6260, 62601V.

166. Snitkovskii, Yu. P. 2001. Silicon surface cleaning by wet etching for IC production in a closed manufacturing system. *Russian Microelectronics* 30(3), 191–194.

167. Kiryushina, I. V., Krasavina, L. Z., Prosiy, A. D., Selivanova, I. N., and Yasnov, V. S. 2004. Cleaning processes of silicon wafers in modified ammonia-peroxide and salt-peroxide solutions. *Izvestiya vysshikh uchebnykh zavedenii. Elektronika* 1, 53–60 (in Russian).

168. Kashkoush, I., Chen, G.-S., Ciari, R., and Novak, R. E. March 4, 2002, filed, and January 4, 2005, published. Cleaning and drying method and apparatus. US Patent 6837944.

169. Gribov, B. G., Lysak, L. V., and Martemyanov, V. S. 2006. A new method for cleaning silicon wafers. *Izvestiya vysshikh uchebnykh zavedenii. Elektronika* 5, 15–19 (in Russian).

170. Kovalev, A. A. 2006. Features of the technique for cleaning semiconductor structures based on electrochemical synthesis and recovery of solutions. *Izvestiya vysshikh uchebnykh zavedenii. Elektronika* 4, 13 (in Russian).

171. Beklemishev, V. I., Levenets, V. V., Makhonin, I. I., Minazhdinov, M. S., and Seletskaya, I. V. 1995. The effect of substrate-surface treatment on the structure of the near-surface Si layer. *Elektronnaya Promishlennost* 2, 54–56 (in Russian).

172. Hirooka, T. and Sakumichi, H. June 22, 1999, filed, and January 4, 2005, published. Cleaning and handling methods of electronic component and cleaning apparatus thereof. US Patent 6837941 B2.

173. Wojtczak, W. A., Seijoo, F., Bernhard, D., and Nguyen, L. March 27, 2001, filed, and June 29, 2004, published. Aqueous cleaning composition containing copper-specific corrosion inhibitor for cleaning inorganic residues on semiconductor substrate. US Patent 6755989.

174. Timoshenkov, S. P., Kalugin, V. V., and Prokopyev Ye, P. 2003. Investigation into the preparation processes of Si wafer surfaces for fabricating SOI structures and microelectronic components. *Mikroelektronika* 32(6), 459–465 (in Russian).

175. Peters, D. W. and Egbe, M. I. February 8, 2002, filed, and September 15, 2004, published. Composition for removing residues from the microstructure of an object. Patent Application EP 1457550 A2.

176. Ilyin, M. K. et al. 1991. Regeneration of freon-based washing compositions using polymer diaphragms. *Elektronnaya tekhnika*, 7th ser. 4, 45–49 (in Russian).

177. Bennett, J. M. 2004. How to clean surfaces. In *35th Annual Boulder Damage Symposium Proceedings: Laser-Induced Damage in Optical Materials 2003.* 5273, p. 195–206.

178. Arnold, N. 2003. Theoretical description of dry laser cleaning. *Applied Surface Science* 208, 15–22.

179. Norton, M. A., Donohue, E. E., Hollingsworth, W. G., McElroy, J. N., and Hackel, R. P. 2004. Growth of laser-initiated damage in fused silica at 527 nm. In *35th Annual Boulder Damage Symposium Proceedings: Laser-Induced Damage in Optical Materials* 2003. 5273, p. 236–243.

180. Kaufman, H. R. 1990. Broad-beam ion sources. *Review of Scientific Instruments* 61(II), 230–236.

181. Kazanskiy, N. L., Kolpakov, V. A., and Kritchevsky, S. V. 2006. Simulating the process of dielectric-substrate surface cleaning in high-voltage gas-discharge plasma. *Proceedings of SPIE* 6260, 62601V-1–62601V-8.

182. Tamura, T. June 6, 2001, filed, and August 3, 2004, published. Plasma cleaning method and placement area protector used in the method. US Patent 6769439.

183. Shandrikov, M. V. 2005. Low-Temperature Plasma Generators Based on Low-Pressure Discharge with Electron Injection from Contracted Arc Discharge. *Author's abstract of the dissertation for the title of candidate of technical sciences,* Institute of High-Current Electronics, Siberian Branch of the Russian Academy of Sciences, Tomsk (in Russian).

184. Dostanko, A. P. et al. 2001. *Plasma Processes in Electronics Production.* Edited by Dostanko, A. P. Minsk: FUAinform (in Russian).

185. Bordusov, S. V. 2002. *SHF Plasma Technology in Electronics Production.* Edited by Dostanko, A. P. Minsk: Bestprint.

186. Bordusov, S. V. 2001. SHF plasma removal of photoconductive coatings from semiconductor wafers. *Journal of Applied Spectroscopy* 68(6), 1026–1029.

187. Danilin, B. S. 1986. Technology- and vacuum-related problems in very-large-scale integration. *Science and Technology Abstracts. Electronics* 8, 133–157. Moscow: VINITI (in Russian).

188. Danilin, B. S. 1982. *Magnetron Sputtering Systems.* Moscow: Radio i svyaz Publisher.

189. Volkov, A. V., Kazanskiy, N. L., and Moiseyev, O. Yu. 2003. Preparation of substrate surface for DOE fabrication using layered photoresist growth. *Computer Optics* 25, 112–117.

190. Ivanovskii, G. V. and Petrov, V. I. 1986. *Plasma and Ion Surface Engineering.* Moscow: Radio i svyaz Publisher (in Russian).

191. Ryazantsev, S. S., Gavrilenko, I. B., and Udalov, Yu. P. 2006. Using the hollow cathode effect for the preparation of dielectric substrates prior to coating. *Fizika i khimiya obrabotki materialov* 2, 132–133 (in Russian).

192. Lazarovich, S. D., Rosenberg, A., Shiloh, J., Statlender, J., and Wurzberg, E. June 22, 2001, filed, and February 3, 2004, published. Plasma treatment of processing gases. US Patent 6685803.

193. Kagadei, V. A. 2002. Surface and Near-Surface Phenomena Occurring in Semiconductors When They Are Exposed to Electron and Hydrogen-Atom Beams. Dr. Sc. (Phys.-Math.) dissertation, Tomsk (in Russian).

194. Volkov, A. V., Kazanskiy, N. L., Kostyuk, G. F., and Pavelyev, V. S. 2002. Dry etching of polycrystalline diamond films. *Optical Memory and Neural Networks (Information Optics)* 11(2), 135–138.

195. Lu, B., Wasson, J. R., Han, S. I., Mangat, P., Golovkina, V., and Cerrina, F. 2003. EUV radiation damage rest on EUVL mask absorber materials. *SPIE Proceedings* 5256, 1232–1238.

196. Farenik, V. I. 2004. Low-pressure high-frequency discharges in low-power-intensive vacuum-plasma etching of microstructures. *Fizicheskaya inzheneriya poverkhnosti* 2(1), 117 (in Russian).

197. Uddin, M. A., Alam, M. O., Chan, Y. C., and Chan, H. P. 2003. Plasma cleaning of the flex substrate for flip-chip bonding with anisotropic conductive adhesive film. *Journal of Electronic Materials* 32(10), 1117–1124.

198. Romanenko, S. V., Kagadei, V. A., Nefeyodtsev, E. V., and Proskurovsky, D. I. 2004. Cleaning of Si and GaAs surface in the atomic hydrogen flow formed by the source based on low-pressure arc discharge. In *The Proceedings of the Seventh International Conference on Modification of Materials with Particle Beams and Plasma Flows.* p. 271–276. Tomsk: V. E. Zuev Institute of Atmospheric Optics at the Russian Academy of Sciences Siberian Branch.

199. Belova, N. G., Valiev, K. A., Lukichev, V. F., and Orlikovskiy, A. A. 1999. Simulation of ion extraction from a high-density plasma source for wide-beam ion implantation. *Russian Microelectronics* 28(5), 316–322.

200. Ogawa, U. and Sato, T. January 10, 2001, filed, and April 27, 2004, published. Plasma processing apparatus. US Patent 6727654.

201. Chernozhukov, N. I. 1955. *Chemistry of Mineral Oils.* Moscow and Leningrad: Gostekhizdat Publisher.

202. Potapov, V. M. 1976. *Stoichiometry: A Textbook.* Moscow: Khimiya Publisher.

203. Cherepnin, N. V. 1973. *Sorption Phenomena in Vacuum Technology.* Moscow: Sovetskoye radio Publisher (in Russian).

204. Blinov, L. M. 1983. Physical properties and applications of Langmuir monomolecular and multimolecular structures. *Russian Chemical Reviews* 52(8), 713–736.

205. Blinov, L. M. 1988. Langmuir films. *Soviet Physics Uspekhi* 31(7), 623–644.

206. Dedkov, G. V. 2000. Nanotribology: Experimental facts and theoretical models. *Physics-Uspekhi* 43(6), 541–573.

207. Mironov, V. L. 2004. *Fundamentals of Scanning Probe Microscopy.* Nizhny Novgorod: Institute for Physics of Microstructures, Russian Academy of Sciences (in Russian).

208. Bulatov, A. N. and Khartov, S. V. 2004. Investigation of air adsorbate on solid substrates through atomic force microscopy. *Izvestiya vysshikh uchebnykh zavedenii. Elektronika* 4, 9–17 (in Russian).

209. Kolpakov, V. A. 2002. Modelling the high-voltage gas-discharge plasma etching of SiO_2. *Russian Microelectronics* 31(6), 366–374.

210. Kazanskiy, N. L., Kolpakov, A. I., and Kolpakov, V. A. 2004. Anisotropic etching of SiO_2 in high-voltage gas-discharge plasmas. *Russian Microelectronics* 33(3), 169–182.

211. Kazanskiy, N. L., Kolpakov, V. A., and Kritchevsky, S. V. 2005. Simulating the process of dielectric-substrate surface cleaning in high-voltage gas-discharge plasma. *Computer Optics* 28, 80–86.

212. Kireev, V. Yu., Nazarov, D. A., and Kuznetsov, V. I. 1986. Ion-stimulated etching. *Elektronnaya Obrabotka Materialov* 6, 40–43 (in Russian).

213. Kireev, V. Yu. and Kremerov, M. A. 1985. Electron-stimulated etching. *Elektronnaya tekhnika. Ser. 3, Microelektronika* 151, 3–12 (in Russian).

214. Brodie, I. and Muray, J. 1985. *Fizicheskiye osnovy mikrotekhnologii.* Edited by Shalnov, A. V. Moscow: Mir Publisher. Originally published as *The Physics of Microfabrication.* 1982. New York: Plenum Press.

215. Anisimov, Yu. N., Galibey, V. I., Ivanchenko, P. A., Kirichenko, I. N., Oleshchyuk, V. I., and Epimakhov, Yu. K. 1987. *Polymerisation Processes and Physicochemical Research Methods.* Kiev: Vishcha shkola (in Russian).

216. Kolpakov, V. A. 2004. Candidate's dissertation. Samara State Aerospace University; Image Processing Systems Institute at the Russian Academy of Sciences, Samara (in Russian).

217. Kazanskiy, N. L., Kolpakov, V. A., and Kritchevsky, S. V. 2006. Simulating the mechanism of electron–ion cleaning for dielectric-substrate surfaces. In *Proceedings of an All-Russian Scientific and Technical Conference Titled Methods for Developing and Researching Materials and Equipment and Economic Aspects of Microelectronics.* p. 58–62. Penza (in Russian).

218. Kazanskiy, N. L., Kolpakov, A. I., Kolpakov, V. A., and Kritchevsky, S. V. 2007. A method for measuring the residual concentration of organic contaminants on a silicon-dioxide surface. In *Proceedings of the Eighth International Scientific and Technical Conference Titled AVIA-2007.* 1, p. 14.5–14.8. Kiev: National Aviation University (in Ukrainian).

219. Kazanskiy, N. L., Kolpakov, V. A., Kolpakov, A. I., Kritchevsky, S. V., and Podlipnov, V. V. 2006. Investigation into cleaning mechanisms of dielectric substrates in gas-discharge plasma. In *Proceedings of an All-Russian Scientific and Technical Conference Titled Current Problems of Radio Electronics and Telecommunications.* Edited by Mironenko, I. G. and Piganov, M. N. Samara: Samara State Aerospace University. p. 88–99 (in Russian).

220. Kolpakov, V. A. 1998. Increasing the contact resistance of subminiature relays for space applications by irradiating their surfaces with an ion-plasma flux. In *Proceedings of BIKAMP-98, the First International Youth Conference on the Future of Information, Space, Aircraft, and Medicine Instrumentation.* p. 38. Saint Petersburg (in Russian).

221. Kolpakov, V. A., Kolpakov, A. I., Maklashov, V. A., and Balakin, A. V. 1998. An instrument for quickly measuring the parameters of semiconductor devices. *Pribory i tekhnika eksperimenta* 6, 142 (in Russian).

222. Seidel, C., Kopf, H., Gotsmann, B., Vieth, T., Fuchs, H., and Reihs, K. 1999. Ar plasma treated and Al metallised polycarbonate: A XPS, mass spectroscopy and SFM study. *Applied Surface Science* 150(1–4), 19–33.

223. Brown, J. D. and Govers, M. R. 1995. Study of titanium–nitrogen films deposited in an electron beam evaporation unit. *Journal of Vacuum Science and Technology A* 13(5), 2328–2335.

224. Spool, A. M. 1994. Studies of adhesion by secondary ion mass spectrometry. *IBM Journal of Research and Development* 38(4), 391–411.

225. Minoru, S., Kazuhiro, H., Shigeru, O., and Akihiro, T. 1995. Scanning force microscope technique for adhesion distribution measurement. *Journal of Vacuum Science and Technology B* 13(2), 350–354.

226. Shcherbina, G. I., Toporov, Yu. P., Akimov, A. V., and Aleynikova, I. N. 1999. An automated probe for determining surface adhesion properties. *Instruments and Experimental Techniques* 42(3), 420–422.

227. Deryagin, B. V., Krotova, N. A., and Smilga, V. P. 1978. *Adhesion of Solids.* New York: Plenum.

228. Chopra, K. D. 1969. *Thin Film Phenomena*. New York: McGraw-Hill.
229. Berstein, V. A., Zaytseva, V. P., Nikitin, V. V., and Zharov, V. A. 1979. On the effect of glow discharge on the glass surface. *Fizika i khimiya obrabotki materialov* 4, 147–150 (in Russian).
230. Kovalenko, V. V. and Upit, G. P. 1983. The effect of the glass-preparation method on the adhesion of vacuum condensates of indium to glass. *Fizika i khimiya obrabotki materialov* 6, 77–80 (in Russian).
231. Kovalenko, V. V. and Varchenya, S. A. 1988. The effect of glow-discharge plasma on metal-condensate adhesion to silicon dioxide and materials based thereon. *Fizika i khimiya obrabotki materialov* 1, 63–68 (in Russian).
232. Kurnosov, A. I. and Yudin, V. V. 1974. *Semiconductor Fabrication Techniques*. Moscow: Vysshaya shkola (in Russian).
233. Gritsina, V. T., Polyakov, N. I., and Poltoratskiy, Yu. B. 1973. A small-sized precision tensile-testing machine. *Zavodskaya laboratoriya* 2, 35–236 (in Russian).
234. Koptenko, V. M. and Kononenko, Yu. G. 1982. Comparing primary hydrocarbon-contaminant sources during vacuum deposition of thin films: Thin-film fabrication and properties. In *Sbornik nauchnykh trudov*. p. 5–12. Kiev: Naukova dumka (in Russian).
235. Danilin, B. S. 1972. *Vacuum Technology in Fabrication of Integrated Circuits*. Moscow: Energiya (in Russian).
236. Moreau, W. M. 1988. *Semiconductor Lithography: Principles, Practices, and Materials*. New York and London: Plenum Press.
237. Samsonov, G. V., ed. 1968. *Handbook of the Physicochemical Properties of the Elements*. New York: Plenum.
238. Semm, B. F. and Goryunov, Yu. V. 1976. *Physicochemical Foundations of Wetting and Spreading*. Moscow: Khimiya.
239. Karakozov, E. S., Kartashkin, B. A., and Shorshonov, M. Kh. 1968. On the kinetics of bond formation during the welding of similar metals in a solid state. *Fizika i khimiya obrabotki materialov* 3, 113–122 (in Russian).
240. Vyatskin, A. Ya. and Makhov, A. F. 1958. Electron deceleration in some metals and semiconductors. *Zhurnal tekhnicheskoi fiziki* 28(4), 740–747 (in Russian).
241. Afanasyev, V. P., Lubchenko, A. V., and Ryzhkov, A. A. 1996. Energy loss of kilo-volt electrons in transmission through layers of a solid. *Poverkhnost* 1, 6–17 (in Russian).
242. Dudko, G. V., Kolegayev, M. A., and Cherednichenko, D. I. 1970. On possible formation and distribution mechanisms of defects in silicon and germanium during electron-beam heating. *Fizika i khimiya obrabotki materialov* 2, 25–29 (in Russian).
243. Seitzm, F. and Turnbull, D., eds. 1956. *Solid State Physics*. London: Academic Books.
244. Vakhidov, Sh. A., ed. 1977. *Radiatively Activated Processes in Silicon*. Tashkent: FAN UzSSR.
245. Polyanin, A. D. and Nazaikinskii, V. E. 2016. *Handbook of Linear Partial Differential Equations for Engineers and Scientists*. London: Taylor & Francis.
246. Novoselova, A. V. et al., eds. 1979. *Physicochemical Properties of Semiconductors: A Reference Book*. Moscow: Nauka Publisher (in Russian).
247. Kazanskiy, N. L., Kolpakov, V. A., Paranin, V. D., and Polikarpov, M. S. 2008. The method of thin metal films adhesion increasing for the lowered dimensions structures. *Proceedings of SPIE* 7025, 70250H-1–70250H-9.
248. Kazanskiy, N. L., Kolpakov, V. A., and Polikarpov, M. S. 2007. The method of thin metal films adhesion increasing for the lowered dimensions structures. In *Proceedings of the Third International Conference Titled Physical Materials Science: Nanomaterials for Technical and Medical Applications*, Edited by Vikarchuk, A. A. p. 353–355. Togliatti: Togliatti State University (in Russian).

249. Kazanskiy, N. L., Kolpakov, V. A., Paranin, V. D., and Polikarpov, M. S. 2007. The method of thin metal films adhesion increasing for the lowered dimensions structures. In *The 2007 International Conference on Micro- and Nanoelectronics: A Book of Abstracts.* p. 1–44. Zvenigorod.

250. Valiev, K. A., Makhviladze, T. M., and Sarychev, M. Ye. 1985. Polymer plasma-chemical etching mechanism. *Doklady Akademii Nauk SSSR* 283(2), 366–369 (in Russian).

251. Sarychev, M. Ye. 1992. Nonlinear diffusion model of polymer resist plasma-chemical etching process: Simulation of technological processes of microelectronics. *Trudy FTIAN* 3, 74–84. Moscow: Nauka Publisher (in Russian).

252. Kireev, V. Yu., Danilin, B. S., and Kuznetsov, V. I. 1983. *Plasmo-Chemical and Ion-Chemical Etching of Microstructures.* Moscow: Radio i svyaz Publisher (in Russian).

253. Bagriy, I. P. and Chechko, G. A. 1989. Simulating Plasma-Chemical Etching Processes in Fabrication of Integrated Circuits. Preprint, V. M. Glushkov Institute of Cybernetics at the National Academy of Sciences of Ukraine, Kiev (in Russian).

254. Kazanskiy, N. L., Kolpakov, V. A., Kritchevsky, S. V., and Ivliev, N. A. 2006. Investigation into the interaction mechanism of dielectric substrates in the device for measuring substrate-surface cleanliness. In *Proceedings of an All-Russian Scientific and Technical Conference Titled Methods for Developing and Researching Materials and Equipment and Economic Aspects of Microelectronics*, p. 32–35. Penza (in Russian).

255. Kazanskiy, N. L., Kolpakov, V. A., Kolpakov, A. I., and Kritchevsky, S. V. 2006. Nondestructive diagnosis of dielectric-substrate surface cleanliness. In *Proceedings of the Seventh International Scientific and Technical Conference Titled AVIA-2006.* 1, p. 11.65–11.68. Kiev: National Aviation University (in Russian).

256. Kazanskiy, N. L., Kolpakov, V. A., Kritchevsky, S. V., and Podlipnov, V. V. 2006. A dynamic evaporator of complex materials. In *Proceedings of an All-Russian Scientific and Technical Conference Titled Methods for Developing and Researching Materials and Equipment and Economic Aspects of Microelectronics.* p. 25–27. Penza (in Russian).

257. Soifer, V. A., Kazanskiy, N. L., Kolpakov, V. A., Kolpakov, A. I., and Podlipnov, V. V. April 4, 2007, filed, and March 10, 2009 published. Evaporator of multicomponent solutions. Patent RU 2348738 (in Russian).

258. Flamm, D. L. 1979. Measurements and mechanisms of etchant production during the plasma oxidation of CF_4 and C_2F_6. *Solid State Technology* 22(4), 109–116.

259. Kazanskiy, N. L., Kolpakov, V. A., Kolpakov, A. I., and Kritchevsky, S. V. 2002. Simulation of ion-plasma cleaning for dielectric-substrate surfaces. In *Proceedings of the Second International Symposium on Aerospace Engineering.* p. 89–90. Saint Petersburg (in Russian).

260. Gerlach-Meyer, V. 1981. Ion-enhanced gas–surface reactions: A kinetic model for the etching mechanism. *Surface Science* 103(213), 524–534.

261. Brown, S. 1961. *Elementary Processes in Gas-Discharge Plasma.* Moscow: Gosatomizdat Publisher.

262. Yasunori, O., Hitoshi, M., and Hiroharu, F. 1996. Spatial structure of electrons and fluorine atoms in a CF_4 RF magnetron plasma. *Plasma Sources Science and Technology* 5(2), 344–348.

263. Daiyu, H., Masahiko, N., Noriharu, T., Koichi, S., and Kiyoshi, K. 1999. Role of reaction products in F^- production in low-pressure, high-density CF_4 plasmas. *Japanese Journal of Applied Physics* 38(10), 6084–6089.

264. Kouji, K., Takashi, K., Takao, I., and Kazuyuki, O. 2001. Spatial structure of electronegative Ar/CF_4 plasmas in capacitive RF discharges. *Japanese Journal of Applied Physics* 40(10), 6115–6116.

265. Kalitkin, N. N. 1978. *Numerical Methods.* Moscow: Nauka Publisher (in Russian).

266. Myshkina, V. V. and Kolpakov, V. A. 1996. Using the conformal-mapping method for calculating the distribution of the electrostatic field in high-voltage gas-discharge devices. In *Abstracts of an International Seminar Titled Differential Equations and Their Applications*. p. 32. Samara (in Russian).

267. Lukichev, V. F. and Yunkin, V. A. 1999. Scaling of silicon trench etch rates and profiles in plasma etching. *Microelectronic Engineering* 46(1–4), 315–318.

268. Coburn, J. W., Winters, H. F., and Chuang, C. J. 1977. Ion–surface interactions in plasma etching. *Journal of Applied Physics* 48(8), 3532–3540.

269. Coburn, J. W. and Winters, H. F. 1979. Ion- and electron-assisted gas surface chemistry: An important effect in plasma etching. *Journal of Applied Physics* 50(5), 3189–3196.

270. Kazanskiy, N. and Kolpakov, V. 2003. Simulation of technological process by etching of microstructures in high-voltage gas-discharge plasma. In *Abstracts of Micro- and Nanoelectronics 2003, An International Conference*. p. 1–53. Zvenigorod (in Russian).

271. Kazanskiy, N. and Kolpakov, V. 2004. Simulation of technological process of microstructures etching in high-voltage gas-discharge plasma. *Proceedings of SPIE* 5401, 648–654.

272. Volkov, A. V., Kazanskiy, N. L., and Kolpakov, V. A. 2001. Simulation of the plasma-chemical etching process for microstructures on quartz substrates. In *Proceedings of an All-Russian Scientific and Technical Conference Titled Micro- and Nanoelectronics 2001*. p. 198–199. Zvenigorod (in Russian).

273. Abroyan, I. A., Andronov, A. N., and Titov, A. I. 1984. *Physical Fundamentals of Electron and Ion Technology*. Moscow: Vysshaya shkola Publisher.

274. Volkov, A. V., Kazanskiy, N. L., and Kolpakov, V. A. 2001. Calculation of the velocity of plasma-chemical etching of quartz. *Computer Optics* 21, 121–125 (in Russian).

275. Harsberger, W. R. and Porter, R. A. 1979. Spectroscopic analysis of RF plasmas. *Solid State Technology* 22(4), 90–103.

276. Horiike, Y. 1983. Dry etching: An overview. *Semiconductor Technologies, Japan Annual Reviews in Electronics, Computers and Telecommunications* 8, 55–72.

277. Poulsen, R. G. and Brochu, M. 1976. Importance of temperature and temperature control in plasma etching. In *Proceedings of the Symposium on Etching for Pattern Definition*. Princeton, NJ: Electrochemical Society.

278. Tikhonov, A. N. and Arsenin, V. Ya. 1977. *Solutions of Ill-Posed Problems*. New York: Halsted.

279. Alifanov, O. M. 1983. Methods of solving ill-posed inverse problems. *Journal of Engineering Physics* 45(5), 1237–1245.

280. Vabishchevich, P. N. and Pulatov, P. A. 1986. Numerical solution of an inverse heat-conduction boundary problem. *Journal of Engineering Physics* 51(3), 1097–1100.

281. Samarskii, A. A. and Vabishchevich, P. N. 1996. *Computational Heat Transfer*. Chichester: Wiley.

282. Kazanskiy, N. L., Kolpakov, A. I., Kolpakov, V. A., and Paranin, V. D. 2007. Temperature measurement of a surface exposed to a low-temperature plasma flux. *Technical Physics* 52(12), 1552–1556.

283. Alifanov, O. M. 1994. *Inverse Heat Transfer Problem*. New York: Springer.

284. Carslaw, H. S. and Jaeger, J. C. 1956. *Conduction of Heat in Solids*. Oxford: Clarendon Press.

285. Tikhonov, A. N. and Samarskii, A. A. 1964. *Equations of Mathematical Physics*. Oxford: Pergamon.

286. Ditkin, V. A. and Prudnikov, A. P. 1966. *Integral Transforms and Operational Calculus*. Oxford: Pergamon.

287. Kikoin, I. K., ed. 1976. *Tables of Physical Quantities*. Moscow: Atomizdat Publisher (in Russian).

288. Malkovich, R. Sh. 2002. Alternative analytical solutions of the diffusion (thermal conductivity) equation for an arbitrary initial concentration (temperature) distribution. *Technical Physics Letters* 28(11), 923–924.

289. Kartashov, E. M. 2001. *Analytical Methods in the Theory of Heat Conduction in Solids.* Moscow: Vysshaya shkola Publisher (in Russian).

290. Soifer, V. A., Kazanskiy, N. L., Kolpakov, V. A., Kolpakov, A. I., and Paranin, V. D. June 14, 2005, filed, and July 10, 2008, published. A method for measuring the surface temperature of a sample irradiated by gas plasma. Patent RU 2328707 (in Russian).

291. Kolpakov, V. A. 2006. Studying an adhesion mechanism in metal–dielectric structures following the surface ion–electron bombardment. Part 1. Modelling an adhesion-enhancement mechanism. *Fizika i khimiya obrabotki materialov* 5, 41–48 (in Russian).

292. Kolpakov, V. A. 2007. Studying an adhesion mechanism in metal–dielectric structures following the surface ion–electron Bombardment. Part 2. The effect of bombardment parameters on adhesion. *Fizika i khimiya obrabotki materialov* 1, 53–58 (in Russian).

293. Popov, V. K. 1967. Electron flux–substance interaction. *Fizika i khimiya obrabotki materialov* 4, 11–24 (in Russian).

294. Bartenev, G. M. and Barteneva, A. G. 1992. *Relaxation Properties of Polymers.* Moscow: Khimiya Publisher.

295. Bechstedt, F. and Enderlein, R. 1988. *Semiconductor Surfaces and Interfaces.* Berlin: Akademie-Verlag.

296. Kazanskiy, N. L. and Kolpakov, V. A. 2009. Effect of bulk modification of polymers in a directional low-temperature plasma flow. *Technical Physics* 54(9), 1284–1289.

297. Mirkin, L. I. 1961. *Handbook of X-Ray Analysis of Polycrystalline Materials.* Moscow: Fizmatgiz Publisher.

298. Volkov, A. V., Kazanskiy, N. L., Moiseyev, O. Yu., and Soifer, V. A. March 27, 2001, filed, and January 20, 2003, published. A method of fabricating diffractive optical elements on diamond and diamond-like films. Patent RU 2197006 (in Russian).

299. Frenkel, Ya. I. 1975. *Kinetic Theory of Liquids.* Leningrad: Nauka Publisher (in Russian).

300. Frenkel, Ya. I. 1972. *Introduction to the Theory of Metals.* Leningrad: Nauka Publisher (in Russian).

301. Kireev, P. S. 1975. *Semiconductor Physics.* Moscow: Vysshaya shkola Publisher (in Russian).

302. Yudin, V. V. 1977. Microdoping of silicon with the aid of electron-beam heating. *Elektronnaya obrabotka materialov* 3, 27–30 (in Russian).

303. Boltaks, B. I. 1972. *Diffusion and Point Defects in Semiconductors.* Leningrad: Nauka Publisher (in Russian).

304. Vavilov, V. S., Kiv, A. Ye., and Niyazova, O. P. 1981. *Mechanisms of Defect Formation and Migration in Semiconductors.* Moscow: Nauka Publisher (in Russian).

305. Valiev, K. A. and Rakov, V. A. 1984. *Physical Principles of Submicron Lithography in Microelectronics.* Moscow: Radio i svyaz Publisher (in Russian).

306. Nikolsky, B. P., ed. 1966. *The Chemist's Reference Book. 1,* Moscow: Khimiya Publisher (in Russian).

307. Vvedenskii, D. A., ed. 1962. *Physical Encyclopaedic Dictionary. 2.* Moscow: Sovetskaya entsiklopediya Publisher.

308. Komov, A. N., Kolpakov, A. I., and Rafayevich, B. D. 1979. Increasing contact conductivity for power semiconductors. *Elektronnaya tekhnika,* ser. 7, 5, 7–10 (in Russian).

309. Kazanskiy, N. L., Kolpakov, A. I., and Kolpakov, V. A. 2002. Studies into a mechanism of catalytic mask generation in irradiation of an Al-Si structure with high-voltage gas-discharge particles. *Computer Optics* 24, 84–90 (in Russian).

310. Maslov, A. A. 1970. *Technology and Design of Semiconductor Devices.* Moscow: Energiya Publisher (in Russian).
311. Tikhonov, A. N. and Samarskii, A. A. 1963. *Equations of Mathematical Physics.* New York: Pergamon Press.
312. Marchuk, G. I. and Shaydurov, V. V. 1979. *Increasing the Accuracy of Solutions to Difference Schemes.* Moscow: Nauka Publisher.
313. Samarskii, A. A. 2001. *The Theory of Difference Schemes.* New York: Marcel Dekker, Inc.
314. Kazanskiy, N. L., Kolpakov, V. A., and Kolpakov, A. I. 2005. Studies into a mechanism of catalytic mask generation in irradiation of an Al-Si structure with high-voltage gas-discharge particles. *Optical Memory and Neural Networks* 14(3), 151–159.
315. Kolpakov, A. I., Kazanskiy, N. L., and Kolpakov, V. A. 2001. Studies into a mechanism of catalytic mask generation in irradiation of an Al-Si structure with high-voltage gas-discharge Particles. In *Proceedings of an International Conference on Mathematical Modelling.* p. 133–135. Samara (in Russian).

Index

209

Milton Keynes UK
Ingram Content Group UK Ltd.
UKHW040101071024
449327UK00019B/731